Global Reconstruction and the Geomagnetic Field during the Palaeozoic

ADVANCES IN EARTH AND PLANETARY SCIENCES

Advances in Earth and Planetary Sciences 10

Supplement Issue to Journal of Geomagnetism and Geoelectricity

Global Reconstruction and the Geomagnetic Field during the Palaeozoic

Proceedings of IUGG Symposium "Global Reconstruction and the Geomagnetic Field during the Palaeozoic" Canberra, 1979, December

Edited by

M.W. McElhinny
A.N. Khramov
M. Ozima
D.A. Valencio

Center for Academic Publications Japan/Tokyo
D. Reidel Publishing Company/Dordrecht · Boston · London

Library of Congress Cataloging in Publication Data

DATA APPEARS ON SEPARATE CARD

ISBN 90-277-1231-X

Published by Center for Academic Publications Japan, Tokyo, in co-publication with D. Reidel Publishing Company, P.O. Box 17, 3300 AA Dordrecht.

Sold and distributed in Japan, China, Korea, Taiwan, Indonesia, Cambodia, Laos, Malaysia, Philippines, Thailand, Vietnam, Burma, Pakistan, India, Bangla Desh, Sri Lanka by Center for Academic Publications Japan, 4-16, Yayoi 2-chome, Bunkyo-ku, Tokyo 113, Japan.

Sold and distributed in the U.S.A. and Canada by Kluwer Boston Inc., 190 Old Derby Street, Hingham, MA 02043, U.S.A.

Sold and distributed in all other countries by Kluwer Academic Publishers Group, P.O. Box 322, 3300 AH Dordrecht, Holland.

Printed in Japan

Preface

Physical and chemical studies of the earth and planets along with their surroundings are now developing very rapidly. As these studies are of essentially international character, many international conferences, symposia, seminars and workshops are held every year. To publish proceedings of these meetings is of course important for tracing development of various disciplines of earth and planetary sciences though publishing is getting fast to be an expensive business.

It is my pleasure to learn that the Center for Academic Publications Japan and the Japan Scientific Societies Press have agreed to undertake the publication of a series "Advances in Earth and Planetary Sciences" which should certainly become an important medium for conveying achievements of various meetings to the academic as well as non-academic scientific communities. It is planned to publish the series mostly on the basis of proceedings that appear in the Journal of Geomagnetism and Geoelectricity edited by the Society of Terrestrial Magnetism and Electricity of Japan, the Journal of Physics of the Earth by the Seismological Society of Japan and the Volcanological Society of Japan, and the Geochemical Journal by the Geochemical Society of Japan, although occasional volumes of the series will include independent proceedings.

Selection of meetings, of which the proceedings will be included in the series, will be made by the Editorial Committee for which I have the honour to work as the General Editor. I and the members of the Editorial Committee will certainly welcome any suggestions that will promote the series. Whenever the convener of a meeting related to earth and planetary sciences is in a position to have to look for a medium for publishing the proceedings please contact us.

Tsuneji Rikitake
General Editor

Preface to Volume 10

This Symposium was held under the joint sponsorship of IAGA, ICG and IASPEI at the IUGG General Assembly held in Canberra from 3–15 December, 1979. The symposium also celebrated the 25th Anniversary of Palaeomagnetic Studies in Australia.

In 1954 Dr. E. Irving came to Australia from Great Britain to set up the first palaeomagnetic laboratory in that region at the Australian National University. Many of the quite fundamental discoveries relating to continental drift and reversals of the earth's magnetic field were made in that laboratory over his ten year stay in Australia. The present laboratory at A.N.U. is still the one he designed but never in fact occupied since it was completed at the time of his departure to Canada.

The main thrust of the symposium was therefore quite appropriately related to the determination of the global distribution of continents during the Palaeozoic from palaeomagnetic data. The analysis of palaeomagnetic data in terms of reversal history during the Palaeozoic was also a subsidiary aim to the symposium. Appropriately the first paper was given by Irving on the analysis of late Palaeozoic data from Europe and North America. Papers by Van der Voo, Smith *et al.* and Scotese *et al.* followed and a lively discussion ensued. Apparently different conclusions were being drawn in detail from the examination and synthesis of the same data. The difference in approach turned out to be essentially a philosophical one. Should one be allowed as an extra variable in these syntheses to suppose that magnetization ages were not precisely the same as the known stratigraphic or isotopic ages?

Particularly appropriate was the presence of Dr. A. N. Khramov from the Soviet Union who gave papers on continental drift of the Siberian and Russian platforms and on reversal history during the Palaeozoic. Khramov has the distinction of writing the first book on palaeomagnetism published in 1958. Irving had this translated at A.N.U. in 1960 and copies of this were made freely available at the Symposium. Irving of course later wrote the first book in English on palaeomagnetism published in 1964.

Not all the papers presented at the Symposium appear in this issue, since many have been published elsewhere. However, the selection of papers presented here form a fitting record of the Symposium and the celebration of 25 years of palaeomagnetic research in Australia.

M.W. McElhinny
September 1980

Papers Presented at IUGG Symposium "Global Reconstruction and the Geomagnetic Field during the Palaeozoic" Canberra, 1979, December 10–11

Session 1. Palaeozoic Maps
Chairman: M. W. McElhinny

The Concept of a Mobile Pangea
P. Morel and E. Irving

Palaeozoic Assembly of Pangea: A New Plate Tectonic Model
R. van der Voo

Palaeozoic World Maps
A. G. Smith, A. M. Hurley, and J. C. Briden

A Model of Continental Drift Since the Cambrian
V. Bucha

Session 2. Early Palaeozoic, Northern Continents
Chairman: E. Irving

Palaeomagnetism and Reconstructions of the Locations of the Siberian and Russian Platforms during the Late Proterozoic and Palaeozoic
A. N. Khramov and V. P. Rodionov

Appalachian Development in Light of Palaeomagnetically Derived and Geologically Interpreted Plate Motions during the Early Palaeozoic
R. T. Haworth

Magnetism of the Mid-Ordovician Tramore Volcanics, SE Ireland and the Question of a Wide Proto-Atlantic Ocean
E. R. Deutsch

Palaeomagnetism of Fossiliferous Lower Cambrian Sediments: New Results from the Tommotion Stage of the Siberian Platform
J. L. Kirschvink and A. Yu. Rozanov

Session 3. Early Palaeozoic Gondwana
Chairman: A. N. Khramov

Date of Formation of Gondwanaland
M. O. McWilliams, M. W. McElhinny, and A. Kroner

Cambrian Palaeomagnetic Data from Southern and Central Australia: Implications for the Unity of the Australian Plate and Gondwanaland
C. T. Klootwyk

Palaeomagnetism of Two Sequences of Lower Ordovician and Upper Precambrian Rocks from Argentina
D. A. Valencio

Palaeomagnetism of the Lower Palaeozoic of Southeast Australia
B. Goleby

Session 4. Late Palaeozoic Reconstructions
Chairman: D. A. Valencio

Palaeomagnetism of the Tibetan Series, Nepal Himalaya: Implication for the Northern Extent of Greater India
C. T. Klootwyk and D. K. Bingham

Polar Wander Path of Central Iran and Its Interpretation
H. C. Soffel and H. G. Foerster

Session 5. Late Palaeozoic Magnetic Field
Chairman: M. Ozima

The Geomagnetic Field in Palaeozoic Times
A. N. Khramov and V. P. Rodionov

Late Palaeozoic Reversals and Excursions of the Geomagnetic Field as Defined by Palaeomagnetic Data from Red Beds and Igneous Rocks from Argentina
D. A. Valencio

Permian Palaeointensities: A Preliminary Study
W. E. Senanayake

In Search of the Illawarra Reversal
B. J. J. Embleton

A Computor Animation of Continental Drift
C. R. Scotese

CONTENTS

Magnetostratigraphy in the Sydney Basin, Southeastern Australia

B. J. J. EMBLETON* and K. L. McDONNELL**

**CSIRO Division of Mineral Physics, North Ryde, NSW, Australia*
***School of Earth Sciences, Macquarie University, North Ryde, NSW, Australia*

(Received January 29, 1980 ; Revised June 23, 1980)

Late Permian and Early Triassic sediments of the Narrabeen Group have been sampled in the coastal section of the Sydney Basin between Wybung Head (the top of the Newcastle Coal Measures to the north of Sydney) and Coal Cliff (above the Bulli Seam south of Sydney). An ancient magnetisation (presumed original) has been isolated only from the red claystone component of the entire section. The magnetisation of the green and buff coloured medium/coarse grained sandstones belonging to the Munmorah Conglomerate and the Tuggerah Formation has been reset during the Late Mesozoic period of magnetic overprinting. This event is primarily responsible for frustrating an attempt to describe properly the magnetostratigraphic record for the Palaeozoic/Mesozoic boundary from deposits in the eastern part of the Sydney Basin. The primary magnetisation of the red beds has survived magnetic reprinting but evidence of overprinting is present during thermal cleaning to 600°C. The detailed magnetostratigraphy of the red beds was revealed after thermal treatment at 650°C and 660°C. The high temperature directions yield a palaeomagnetic south pole for the Patonga Claystone at latitude 30.4°S, longitude 146.9°E, $\alpha_{95}=7.8°$ and the difference vectors calculated from the low temperature directions yield an overprint south pole position at latitude 57.3°S, longitude 143.4°E, $\alpha_{95}=3.6°$.

In the light of recent palaeomagnetic investigations in southeastern Australia and the recognition of ubiquitous Late Mesozoic magnetic overprinting, a reassessment of earlier results based upon samples collected at least in the eastern regions of the Sydney Basin may be required.

1. Introduction

The Sydney Basin in southeastern Australia is a major structural unit within which more than 4,800 m of marine and fluvial sediments were deposited, mainly between the Late Carboniferous and the Middle Triassic. It contains the major Permian coal deposits in New South Wales (BRANAGAN *et al.*, 1976). Regional stratigraphic correlation, especially in the fluvial parts of the sequence, is severely hampered by the time transgressive nature of basinal lithofacies and the scarcity of continuous and reliable time parallel horizons. These factors have hindered our understanding of the interrelationships of the sedimentary units and attempts to reconstruct the depositional history of the basin.

Since the geomagnetic field of the Late Palaeozoic is characterised by a long period of reversed polarity (IRVING and PARRY, 1963; KHRAMOV, 1967), identification of the transition which marks the beginning of the Early Triassic mixed interval (e.g. HELSLEY, 1969) will potentially provide a time horizon suitable for intra- and inter-basin correlation. The term Illawarra Reversal was proposed by IRVING and PARRY (1963) for the transition which they originally predicted upon the basis of early Triassic normal magnetisation (IRVING, 1963) following the late Palaeozoic reversed magnetisation measured in igneous and sedimentary rock units within the Sydney Basin.

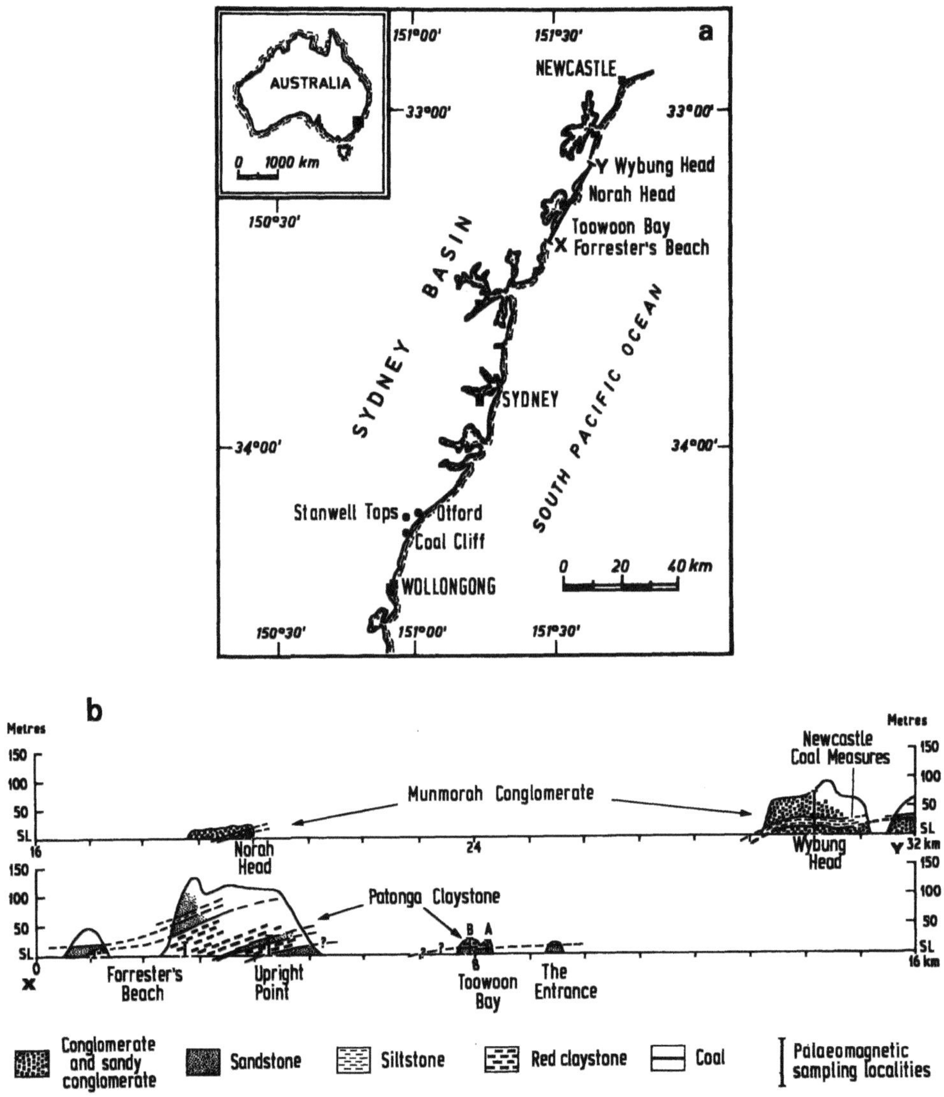

Fig. 1. a) Locality map. Coastal section XY is shown in Fig. 1b at an enlarged scale. b) Coastal section XY showing palaeomagnetic sampling localities. Total length of section 32 km $V:H = 10:1$.

A project has therefore been commenced to investigate the magnetostratigraphy of sections within the basin with a view to establishing correlation guidelines based on the geomagnetic field reversal stratigraphy. Detailed sampling within the Late Permian-Middle Triassic Narrabeen Group has been carried out north and south of Sydney at coastal outcrops between Wybung Head and Coal Cliff (Fig. 1). The beds are essentially flat lying, regional dips of 1 to 2° probably being the result of Tertiary warping.

The lower units of the Narrabeen Group in the coastal section north of Sydney are characterised by conglomerates and coarse grey, green, and buff coloured sandstones. The sediments were carried by streams and deposited in channels located between flood plain fine sands and thin bands of red claystone formed in part by overbank deposition. Higher in the section a relatively persistent red claystone horizon is encountered; in the northern coastal section this is known as the Patonga Claystone and a similar horizon (though possibly younger) south of Sydney is the Bald Hill Claystone (HERBERT, 1970).

2. Reversal Stratigraphy

Stable magnetic remanence measured in all sedimentary types other than the red claystones did not survive thermal treatment above 300°C. The demagnetisation paths in the vector diagrams (Fig. 2) demonstrate this behaviour. Clearly, the elucidation of characteristic primary components of magnetic remanence in these sediments, if retained, will depend upon combined application of suitable cleaning methods; this is to be investigated.

In contrast, the magnetic remanence measured in red claystones was stable to thermal demagnetisation to 660°C. The detailed reversal stratigraphy was revealed only after treatment above 600°C. The vector demagnetisation diagrams (Fig. 2) show the decay of two principal components. The high temperature directions yield the reversal patterns shown in Fig. 3. The pattern obtained is independent of diagenetic features and the transitions do not coincide with lithological boundaries. The metre scales for each section show the thickness of section sampled in each case.

The lowest red claystone bands encountered are at Norah Head. They occur between 200 and 300 m below the base of the Patonga Claystone and are interbedded with the higher beds of the basal unit, the Munmorah Conglomerate, of the Narrabeen Group. The Permian-Triassic boundary probably lies within the Munmorah Conglomerate (HELBY, 1973). The transition from reversed to normal polarity represents the oldest reversal encountered.

The base of the Patonga Claystone crops out close to sea-level at Toowoon Bay. The 16 m section here reveals a dominantly reversed to normal polarity pattern and there are two disturbed zones (normal events?) within the lower reversed section. Moving south towards Sydney and higher in the succession, the section labelled Forrester's Beach in Fig. 3 yields a similar reversal stratigraphy. If the base of this section, sampled at Upright Point, lies stratigraphically above the top of the Toowoon Bay section a transition from normal to reversed polarity has been missed. There is a total absence of outcrop between those two localities—approximately 3 km apart. Normal magnetisation persists to the top of the Patonga Claystone, sampled at the south end of Forrester's Beach.

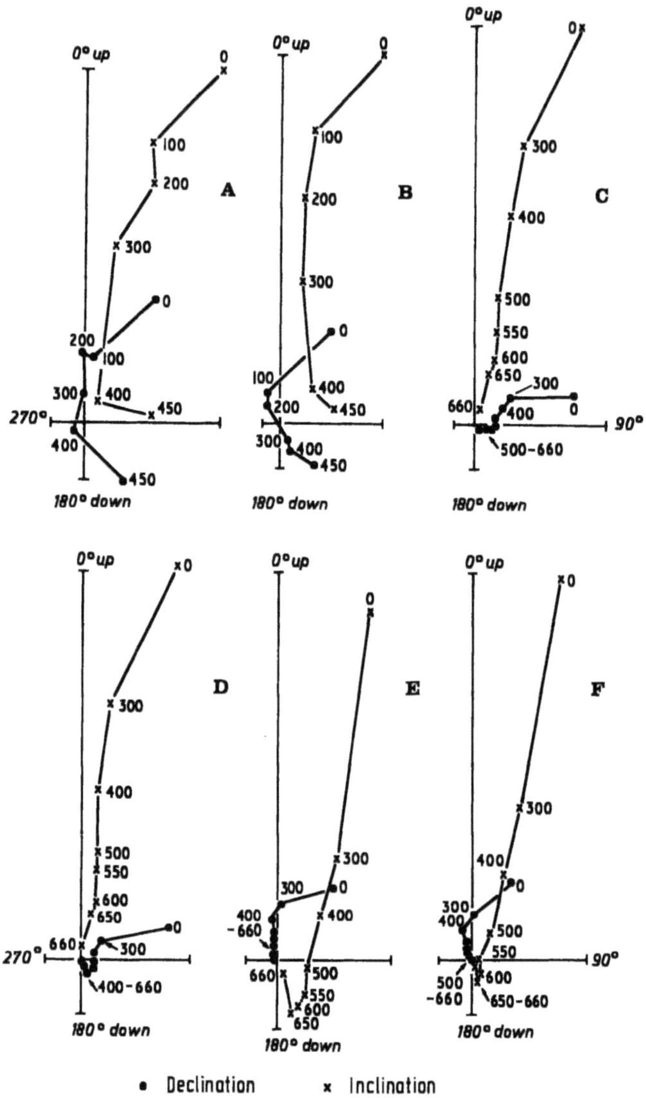

Fig. 2. Vector demagnetisation diagrams. Declination (dots) and inclination (crosses) projected onto the horizontal and vertical planes shown directly at each demagnetisation step. A and B are examples of the intermediate to high temperature magnetic remanence instability in sandstones; C and D examples of primary-normal red claystone; E and F examples of primary-reversed red claystone.

To the south of Sydney, the Bald Hill Claystone was sampled at two road-cut localities about 1 km apart, each covering 3 m of section. The results are summarised in Table 1. In both cases the polarity pattern shows a reversed interval followed by a normal interval.

Although it is not unreasonable to suggest a correlation between the two sections south of Sydney, it remains premature to extend the correlation to the north coast

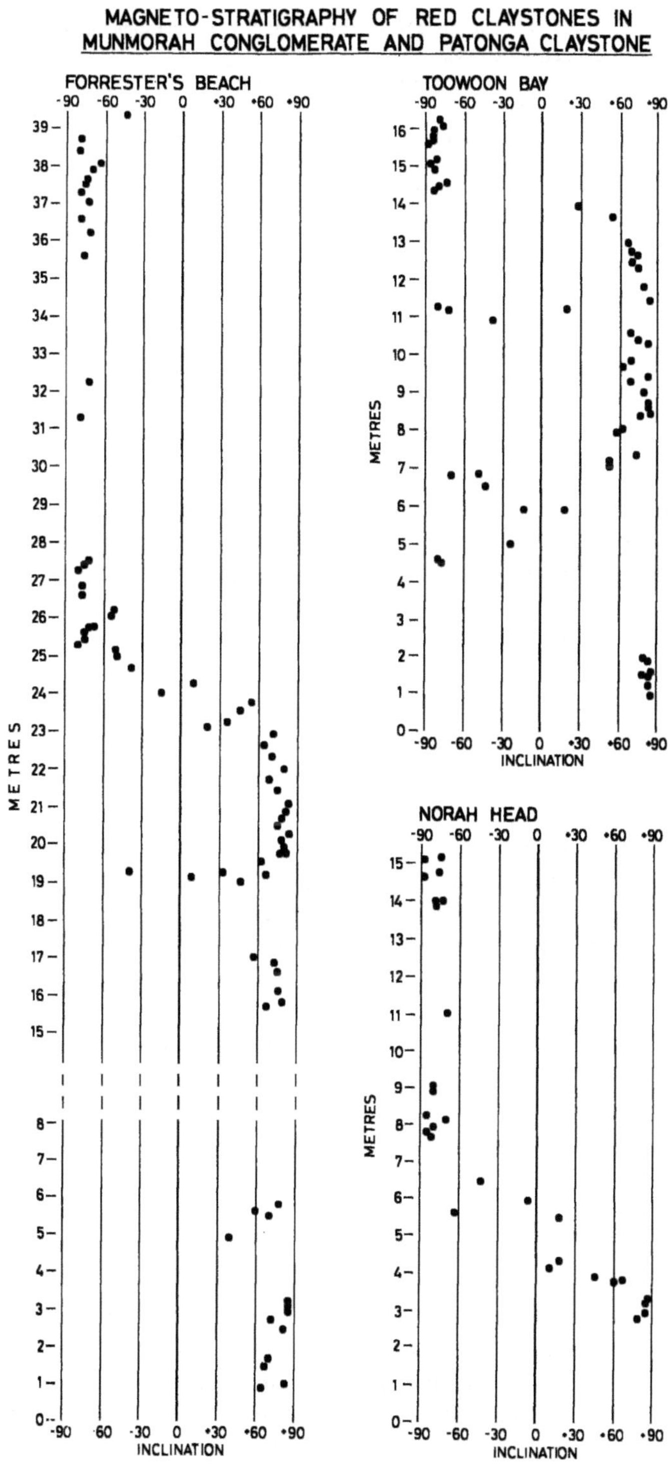

Fig. 3. Magnetic polarity sequences in the red claystones.

Table 1. Bald Hill Claystone reversal stratigraphy. Inclinations measured after treatment of 600°C.

Road-cut locality Arbitrary level	Otford Inclination of magnetisation	Arbitrary level	Stanwell Tops Inclination of magnetisation
3.0 m	−72°	3.0 m	−76°
2.85 m	−86°	1.7 m	−78°
2.5 m	−76°	0.6 m	+69°
2.4 m	−10°	0.45 m	+41°
0.5 m	−15°	0.0 m	+42°
0.3 m	+28°		
0.2 m	+64°		
0.1 m	+69°		
0.0 m	+89°		

sections. The preferred geological viewpoint would be that the Bald Hill Claystone lies stratigraphically above (and is probably younger than) the Patonga Claystone. It is well-known that the Early Triassic is characterised by frequent reversals of the geomagnetic field (e.g. HELSLEY and STEINER, 1974) and in these circumstances the repeated occurrence of a reversed to normal transition is insufficient evidence in itself for establishing correlation.

The behaviour of the geomagnetic field during the transitions has also been investigated. The intensity data have been grouped according to the polarities obtained after high temperature treatment. Because the normal and reversed directions are near-vertical, the transitional averages were estimated from those specimens whose inclinations of magnetisation lay between +60° and −60°. The average intensities (and standard errors of the means) calculated after the removal of the overprint components were: for the normal intervals of the Forrester's Beach and Toowoon Bay sections, $44.05 \pm 5.0\ \mu G$ and $33.0 \pm 7.0\ \mu G$ respectively; for the reversed intervals, $42.0 \pm 5.0\ \mu G$ and $25 \pm 1.0\ \mu G$ respectively; and for the transitional intervals, $3.0 \pm 0.7\ \mu G$ and $4.4 \pm 1.3\ \mu G$ respectively. The intensities are lower in the transitions by one order of magnitude. The data are interpreted to indicate that the polarity intervals result from reversals of the geomagnetic field and that, during the transitions, the magnetising field was correspondingly weak. There is no clear signature within the directional information that is repeated and upon which correlation can be suggested (cf. KHRAMOV and PETROVA, 1972, p. 332).

3. Palaeomagnetic Directions and Poles

Early work on the red claystones (the Narrabeen Chocolate Shales) by IRVING (1963) involved thermal demagnetisation to only 300°C. At that temperature all directions remained normal, and it was on this basis that the Illawarra Reversal was proposed. All 300°C directions from this study (Table 2), except for a 2 m section at Toowoon Bay, also remained normally magnetised. The measured components at this temperature in fact represent a complex multi-component magnetisation comprising a stubborn normal overprint and a weak normal and reversed primary component. At 300°C a present field

Table 2. 300°C directions in the Patonga Claystone and lower red claystones of the Narrabeen Group.

Section	N	Dec°	Inc°	R	α_{95}°
Norah Head	19	014.0	−79.0	18.706	4.3
Toowoon Bay A	7	327.5	+79.0	6.322	21.5
B	39	355.5	−66.0	36.253	13.8
Upright Point	13	006.0	−62.0	12.876	4.2
Forrester's Beach					
North	49	024.5	−74.0	47.916	3.0
South	15	054.5	−79.0	14.352	8.4
Otford	17	352.5	−77.5	16.359	7.2
Mean	6	010.5	−73.5	5.935	7.7
IRVING (1963)	4 (sites)	336.5	−82.0	3.965	10.0

N=number of samples, Dec°=declination of magnetic remanence, Inc°=inclination of magnetic remanence, R=resultant of N unit vectors, α_{95}°=semi-angle of the cone of confidence calculated at the 95% probability level (FISHER, 1953).

viscous remanent magnetisation has possibly been eliminated.

Difference vectors calculated after incremental thermal demagnetisation treatment between 200°C and 550°C revealed the persistent removal of the overprint component in all red claystone specimens from the north coast. These results are summarised in Table 3. The high temperature components of magnetisation with mixed polarities (regarded here as primary) have been analysed with respect to polarity within each section (Table 4). They yield a palaeomagnetic pole for the Early Triassic at latitude 30.4°S, longitude 146.9°E, and $\alpha_{95}=7.8°$. The overprint directions yield a pole position at latitude 57.3°S, longitude 143.4°E, and $\alpha_{95}=3.6°$.

The overprint pole lies on the Late Cretaceous section of the apparent polar wander curve for Australia (EMBLETON, 1980). The severity of magnetic overprinting has hindered earlier attempts to define accurately the configuration of the Mesozoic geomagnetic field from palaeomagnetic studies in southeastern Australia: this problem has been discussed more fully by SCHMIDT (1976a, b).

4. Concluding Remarks

Magnetostratigraphy will no doubt play an important role in establishing correlation guidelines within the Sydney Basin. Clearly this requires the development of palaeomagnetic techniques suitable to the variety of rock types present. In addition, the identification of the Late Cretaceous magnetic overprint has important implications for models concerned with the geological evolution of the basin, particularly in relation to the timing of formation of its rifted eastern margin. The overprint was probably acquired during a period of burial and subsequent uplift caused by the removal of overburden at the time of initial rifting.

The Early Triassic palaeomagnetic pole position is different from the pole reported by IRVING (1963). The reasons for this have been outlined above and consequently we may

Table 3. Difference vectors averaged by section in the red claystones of the Munmorah Conglomerate and the Patonga Claystone. Removed vectors shown to be uniformly directed irrespective of polarity.

Section	Normal magnetisation					Transition					Reversed magnetisation				
	N	Dec°	Inc°	R	α_{95}°	N	Dec°	Inc°	R	α_{95}°	N	Dec°	Inc°	R	α_{95}°
Norah Head	12	013.8	−82.1	11.882	4.5	—	—	—	—	—	4	356.4	−77.0	3.991	4.9
Toowoon Bay	13	006.2	−77.3	12.949	2.6	9	349.6	−77.4	8.959	3.7	17	023.2	−76.2	16.912	2.6
Forrester's Beach North	14	025.0	−76.8	13.808	4.8	7	010.9	−76.8	6.966	4.5	15	013.6	−75.2	14.936	2.5
Otford	7	013.8	−76.5	6.867	9.0	—	—	—	—	—	9	009.0	−73.9	8.829	7.5
Mean (section unit weight)	4	014.8	−78.2	3.996	3.5	—	—	—	—	—	4	010.7	−75.7	3.996	3.3

Mean direction (unit weight to polarity intervals and transition sections):

N	Dec°	Inc°	R	α_{95}°
10	010.2	−77.1	9.986	1.9

See footnotes to Table 2.

Table 4. High temperature directions (660°C) averaged by section for polarity intervals in the red claystone of the Munmorah Conglomerate and the Patonga Claystone.

Section	Normal magnetisation					Reversed magnetisation				
	N	Dec°	Inc°	*R*	α_{95}°	*N*	Dec°	Inc°	*R*	α_{95}°
Norah Head	14	083.4	−88.0	13.825	4.6	—	—	—	—	—
Toowoon Bay	13	017.9	−82.9	12.971	2.0	19	317.0	+82.8	18.707	4.3
Upright Point	—	—	—	—	—	11	341.1	+82.9	10.736	7.5
Forrester's Beach										
North	10	125.0	−83.8	9.921	4.5	14	051.6	+86.1	13.748	5.5
South	13	019.3	−88.8	12.914	3.5	—	—	—	—	—
Mean (section unit weight)	4	064.6	−87.3	3.990	5.3	3	343.9	+84.9	2.994	7.0

Mean direction irrespective of polarity:

N	Dec°	Inc°	*R*	α_{95}°
7	125.5	−87.5	6.974	3.9

See footnotes to Table 2.

need to critically review certain other results obtained previously for the Late Palaeozoic from southeastern Australia. Although field and laboratory stability tests carried out by IRVING (1966) on Carboniferous rocks from New South Wales would seem to have established the position of the Late Carboniferous pole beyond doubt, further work is required to investigate the configuration of the geomagnetic field during Permian time.

The field sampling programme was assisted by Dr. J. Kirschvink, J. Nolan, and B. Taylor. Dr. Kirschvink was also involved with the analysis of results. All measurements were carried out in the palaeomagnetic laboratory of the Research School of Earth Sciences, Australian National University, with a cryogenic magnetometer kindly loaned by Dr. E. M. Shoemaker of the California Institute of Technology.

REFERENCES

BRANAGAN, D., C. HERBERT, and T. LANGFORD-SMITH, *An Outline of the Geology and Geomorphology of the Sydney Basin*, 60 pp., Science Press for the University of Sydney, Sydney, 1976.

EMBLETON, B. J. J., The palaeomagnetism of Australia and Antarctica, Contrib. to Study Group 3, WG10, I.C.G., 1980 (in press).

FISHER, R. A., Dispersion on a sphere, *Proc. R. Soc. Lond.*, **A217**, 295–305, 1953.

HELBY, R., Review of Late Permian and Triassic palynology of New South Wales, *Spec. Publs. Geol. Soc. Aust.*, **4**, 141–155, 1973.

HELSLEY, C. E., Magnetic reversal stratigraphy of the Lower Triassic Moenkopi Formation of Western Colorado, *Geol. Soc. Am. Bull.*, **80**, 2431–2450, 1969.

HELSLEY, C. E. and M. B. STEINER, Paleomagnetism of the Lower Triassic Moenkopi Formation, *Geol. Soc. Am. Bull.*, **85**, 457–464, 1974.

HERBERT, C., A synthesis of Narrabeen Group nomenclature, Sydney Basin, *Geol. Surv. New South Wales Quart. Notes*, October, 3–10, 1970.

IRVING, E., Paleomagnetism of the Narrabeen Chocolate Shales and the Tasmanian Dolerite, *J. Geophys. Res.*, **68**, 2283–2287, 1963.

IRVING, E., Paleomagnetism of some Carboniferous rocks from New South Wales and its relation to geological events, *J. Geophys. Res.*, **71**, 6025–6051, 1966.

IRVING, E. and L. G. PARRY, The magnetism of some Permian rocks from New South Wales, *Geophys. J. R. Astr. Soc.*, **7**, 395–411, 1963.

KHRAMOV, A. N., Magnetic field of the earth in the Late Palaeozoic, *Fiz. Zemli*, **1**, 86–108, 1967.

KHRAMOV, A. N. and G. N. PETROVA, Paleomagnetism with special reference to research in the U.S.S.R., *Tectonophysics*, **13**, 325–340, 1972.

SCHMIDT, P. W., The non-uniqueness of the Australian Mesozoic palaeomagnetic pole position, *Geophys. J. R. Astr. Soc.*, **47**, 285–300, 1976a.

SCHMIDT, P. W., A new palaeomagnetic investigation of Mesozoic igneous rocks in Australia, *Tectonophysics*, **33**, 1–13, 1976b.

Early Palaeozoic Palaeomagnetism in South East Australia

Bruce R. GOLEBY

Research School of Earth Sciences, Australian National University, Canberra A.C.T., Australia

(Received February 25, 1980 ; Revised April 30, 1980)

The palaeomagnetism of 22 formations from the Cowra Trough and Molong High of South East Australia have been studied. Samples of Ordovician through Middle Devonian age were collected in three major areas, covering a time period crucial to the understanding of the Palaeozoic apparent polar wander path of Australia. Five magnetic components have been identified: 1) A recent field overprint, 2) a secondary component related to the break up of Australia and Antarctica, 3) a predominent Devonian secondary overprint associated with the major orogenic episode, 4) a weak Silurian overprint associated with an earlier orogenic episode, and 5) primary magnetic components ranging in age from the Ordovician to the Devonian. The directions obtained enable construction of a well defined polar wander path for South East Australia and Australia as a whole.

1. Introduction

Since the work of LUCK (1973) on the Early Palaeozoic of South East Australia, there have been three interpretations of the available pole positions for Australia. MCELHINNY and EMBLETON (1974) and EMBLETON *et al.* (1974) proposed a two plate tectonic model based on a difference of 90 degrees in the apparent pole path for the South Eastern part of the Tasman Geosyncline (as defined by PACKHAM, 1969) compared with the simplest interpretation of the pole path for the stable central Australian craton (Fig. 1A). Poles SV, MP, AV, and BG represent the Siluro-Devonian track of South East Australian sub-plate. Since the Late Devonian, the sub-plate was welded to the main craton producing a common track for the whole of Australia.

SCHMIDT and MORRIS (1977), by reversing the polarity of the Cambro-Ordovician poles constructed a single apparent polar wander path using all the polar data (Fig. 1B). With SCHMIDT and MORRIS's (1977) model the suggested Early Devonian age of magnetization of the Upper Ordovician to Upper Devonian Mereenie Sandstone (MS), and the very fast polar wander rates required for this interpretation have yet to be explained successfully. Two strong points are in favour, 1) all the reliable Australian poles are used, and fit stratigraphically in sequence. 2) The angular distance between the Ordovician and the Silurian sections requires less than 90 degrees movement.

The third interpretation involves constructing a sequence of three 'loops' (MOREL and IRVING, 1978, path 'Y,' Fig. 1C), one during the Cambro-Ordovician, another during the Siluro-Devonian, and a third during the Late Devonian. The Stairway Sandstone (SS;

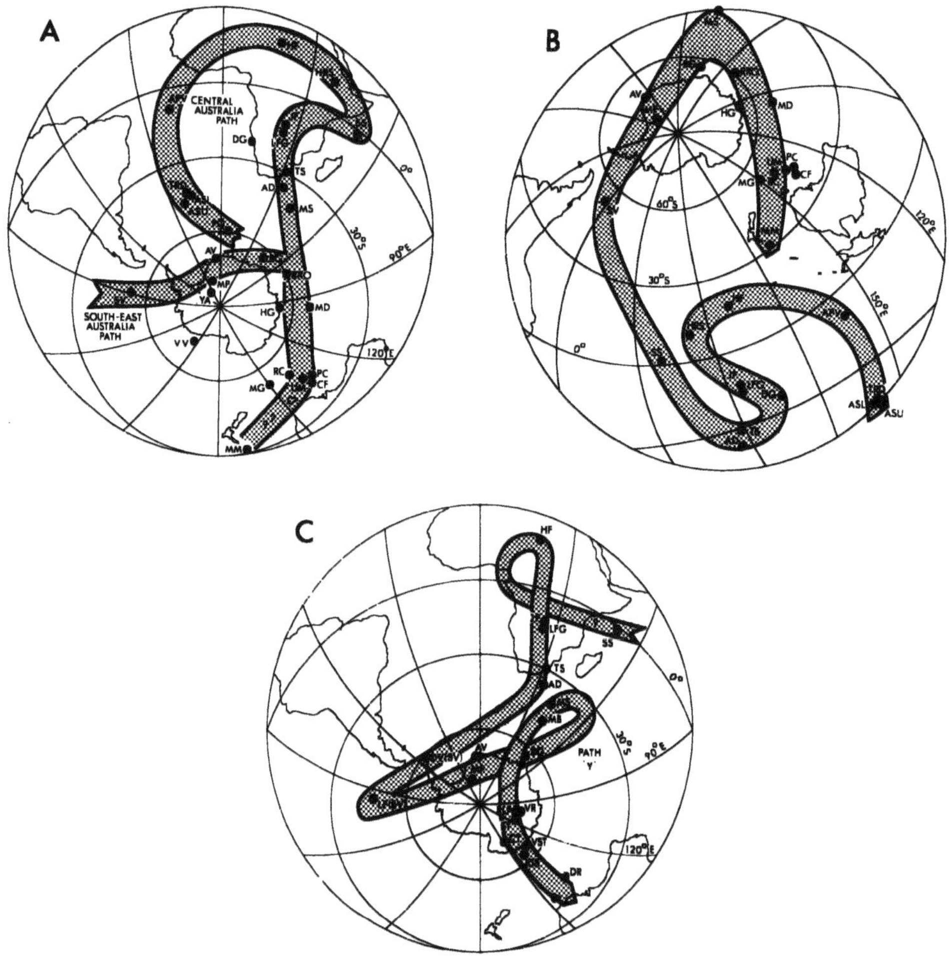

Fig. 1. A) MCELHINNY and EMBLETON's (1974) apparent polar wander path for Australia. The central Australian path defined by Early Palaeozoic poles connects with the South-eastern Australia path during the Late Devonian. B) SCHMIDT and MORRIS's (1977) path after adopting a reversed polarity for pre-Devonian poles. All the Australian poles fit stratigraphically in sequence. C) MOREL and IRVING's (1978) path 'Y'. The connection of the central Australian Ordovician poles to the South-eastern Australian Silurian poles is supported by new results, but the rest of the proposed path is inconsistent, both from stratigraphic relations and the validity of poles used. Polar linear projections. Poles are PQ, Pound Quartzite; ASU and ASL, Upper and Lower Arumbera Sdst; TRD, Todd River Dolomite; APV, Antrim Plateau Volc; HF, Hudson Fm; HRS, Hugh River Shale; SS, Stairway Sdst; JF, Jinduckin Fm; LFG, Lake Frome Group; DG, Dundas Group; TS, Tumblagooda Sdst; AD, Arooma Dam Seds; MS, Mereenie Sdst; SV, Silurian Volc; AV, Ainslie Vols; MP, Mugga Porphyry; BG, Bowning Group; RRO, Ross River Overprint; HG, Housetop Granite; MD, Mulga Downs; RC, Rocky Creek Cong; PC, Permo-carb. Volc; MG, Main Glacial Stage; CF, Currabubulla Fm; UM, Upper Marine Latites; MM, Milton Monzonite; DV(SV), Duro Volcs (Sil Volc); LF(SV), Laidlow Fm (Sil Volc); MB, Murrumbidgee Series; RLF, Redbeds-Lochiel Fm; ST, Tasmanian Sdst; VST, Tasmanian Volc and Seds; KIS, Kangaroo Island Seds; DR, Dotswood Redbeds; CT, Catombal Fm; and VR, Redbeds Vict.

EMBLETON, 1972) is Middle Ordovician (Upper Llanvirnian to Llandeilian; WELLS *et al.*, 1970) and the Hudson Formation (LUCK, 1972) is Middle Cambrian, yet MOREL and IRVING (1978) have their Cambro-Ordovician path passing through the Stairway Sandstone, then the Hudson Formation before connecting with the Cambro-Ordovician Jinduckin Formation (LUCK, 1972), and Lake Frome Group (EMBLETON and GIDDINGS, 1974) poles (Fig. 1C). The Upper Devonian 'loop' is based on selected poles assumed secondary components, yet valid primary poles have been left out and therefore this loop (Fig. 1C) is suspect. Secondly MOREL and IRVING's (1978) path requires movement through greater than 90° angular distance.

The purpose of this paper is to report briefly new palaeomagnetic results obtained from rock sequences of suitable age and location so as to resolve this ambiguity in the Early Palaeozoic polar wander path for Australia. Details on local geology, age relations, rock magnetic properties, and individual results will be presented elsewhere.

2. Geology of the Sampling Areas

Rock samples were collected from three areas within the Tasman Geosyncline of South East Australia. The Tasman Geosyncline has been subdivided into a series of

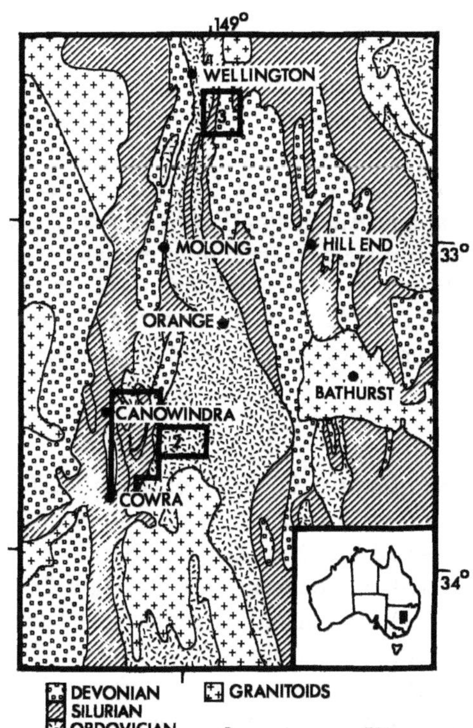

Fig. 2. Regional geology of the areas sampled. The area is part of the Molang High/Cowra Trough region of the Tasman Geosyncline of South-east Australia.

north-south trending tectonic provinces, two of which are the Molong High and Cowra Trough (Fig. 2).

The Molong High is dominantly Ordovician, with localized areas of Silurian and Devonian. The Ordovician sequences represent three episodes of tectonic activity: initial andesitic volcanism during the Early Ordovician, followed by a quiet period enabling the deposition of limestones, siltstones, and arenites during the Middle-Late Ordovician. This was followed by a further period of volcanic activity resulting in volcanic tuffs and lavas. Samples were collected stratigraphically through this sequence, and where possible, in the overlying Siluro-Devonian.

The Cowra Trough (Fig. 2) formed as a result of internal rifting within the Molong High during the Late Ordovician (SCHEIBNER, 1973). It is dominantly Silurian in age, with isolated areas extending into the Middle Devonian.

Four peaks in orogenic activity have been identified within the Tasman Geosyncline during the Early Palaeozoic, each resulting in either localized or widespread deformation at different times in different areas. The major folding in the areas sampled occurred during the Bowning Orogeny (Early Devonian), with minor folding, uplift and erosion occurring during the Early-Middle Silurian (Benambran Orogeny), the Middle Devonian (Tabberabberan Orogeny), and the Early Carboniferous (Kanimblan Orogeny).

3. Experimental Procedures

Samples were collected either as blocks or as drill cores orientated using a modified suncompass (EMBLETON and EDWARDS, 1973). Measurements of intensity and direction were made using either a 'complete results' DIGICO spinner magnetometer (MOLYNEUX, 1971) or a Super Conducting Technology (SCT) cryogenic magnetometer (GOREE and FULLER, 1976). Thermal and Alternating Field (AF) demagnetization procedures were carried out in air and in zero magnetic field using equipment similar to that described by MCELHINNY *et al.* (1971) and MCELHINNY (1966) respectively. Where specimens showed evidence of weathering in the form of a red staining, chemical demagnetization was carried out using a pressurized leaching apparatus (BUREK, 1971).

4. Results

The results will be discussed in terms of a particular time period, and hence, generalizations are made here for each time period. Details regarding each particular formation are to be presented elsewhere. Sampling localities were situated either on both limbs of the dominant fold, and/or from areas having a different bedding orientation. Application of GRAHAM's (1949) fold test on characteristic directions from these samples enabled positive recognization of the pre-folding direction. Each specimen was subjected to stepwise thermal or AF demagnetization up to a maximum of 675°C or 180 nT respectively to analyse the remanence components.

Fig. 3. Compilation of magnetic direction results. For the Ordovician and Silurian each point represents a mean of a minimum of three sites, while each point for the Devonian represents the mean of a single formation. Open circles, negative dips. Equal angle projection.

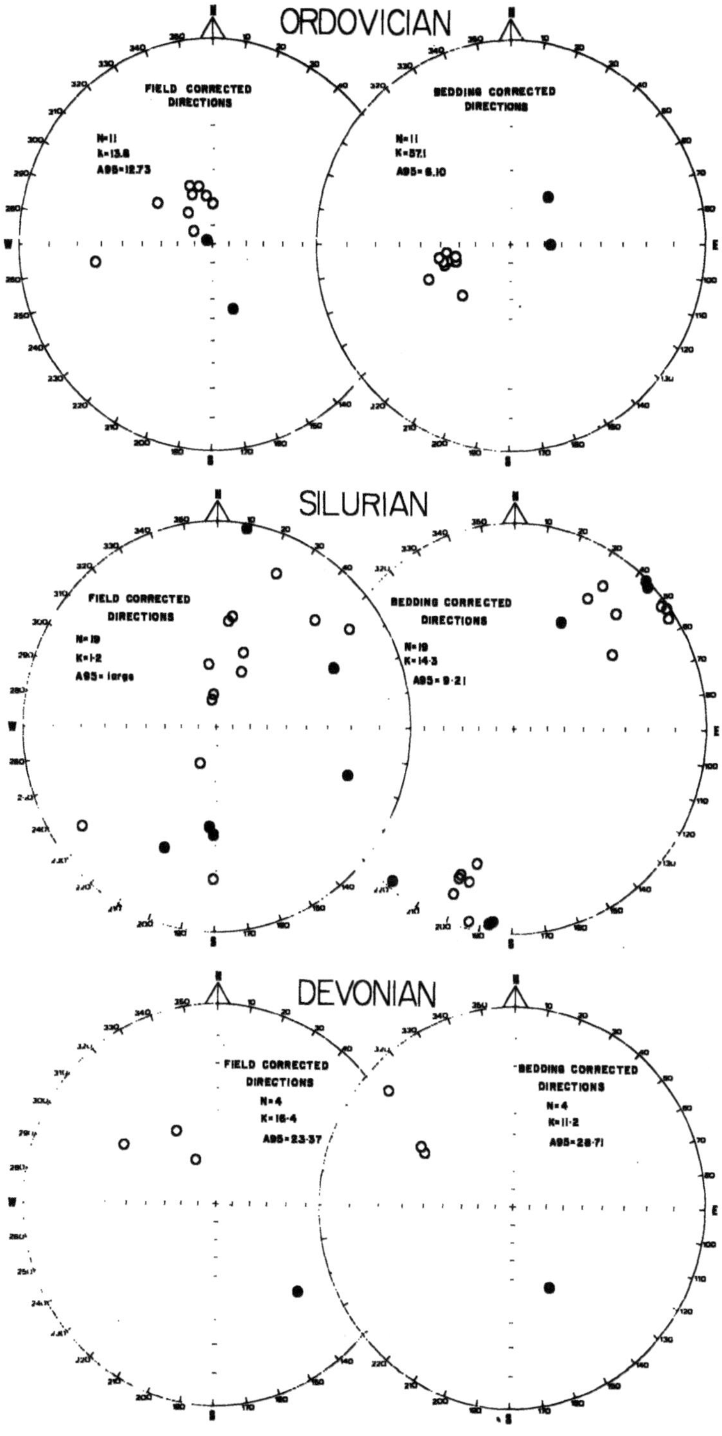
ORDOVICIAN
FIELD CORRECTED DIRECTIONS
N=11
K=13.8
A95=12.73
BEDDING CORRECTED DIRECTIONS
N=11
K=57.1
A95=6.10
SILURIAN
FIELD CORRECTED DIRECTIONS
N=19
K=1.2
A95= large
BEDDING CORRECTED DIRECTIONS
N=19
K=14.3
A95=9.21
DEVONIAN
FIELD CORRECTED DIRECTIONS
N=4
K=16.4
A95=23.37
BEDDING CORRECTED DIRECTIONS
N=4
K=11.2
A95=28.71

Table 1. Mean palaeomagentic data.

Age	Decl	Incl	k	A_{95}	Lat (°N)	Long (°E)	DP	DM
Middle-Late Ord.	251.1	−55.5	57.1	6.10	6.14	19.09	6.22	8.71
Early-Middle Sil.	45.9	−8.3	37.8	8.49	−38.23	34.55	4.31	8.56
Middle-Late Sil.	19.1	10.2	28.6	9.19	−47.34	357.48	4.71	9.30
Early-Middle Dev.	320.9	−24.4	11.2	28.71	−43.96	253.20	17.32	32.10
Secondary components								
Early-Middle Dev.	325.7	−12.8	26.7	9.53	−50.42	272.46	5.01	9.77
Mesozoic	20.6	−76.5	74.9	6.44	−56.62	132.78	11.08	11.94

Re-corrected AV and MP Poles

Formation name	Age	Old pole pos.		New pole pos.		
		Lat (°S)	Long (°E)	Lat (°S)	Long (°E)	
Mugga Porphyry	E. Ludlovian	80.	340.	63.2	7.9	
Ainslie Volc	Late Wenlock	71.	353.	72.3	341.2	for 5° dip
				67.6	285.3	for 35° dip

Eight Ordovician formations were investigated, three of which were completely overprinted. From the remaining Ordovician Formations a primary component and up to four secondary overprint directions were isolated. These were identified using ZIJDERVELD's (1967) orthogonal projections and the least squares method of data analysis. Each overprinting could be related to a major tectonic episode in South East Australia. The primary direction of magnetization was isolated within the temperature range 400°C–650°C. Field and bedding corrected directions for the Ordovician are shown in Fig. 3. After bedding correction the primary directions from the Middle-Late Ordovician Formations clustered about a mean direction of $D=251.1°$, $I=-55.5°$, with Fisher statistics $A_{95}=6.1°$ and $k=57.1$ (Table 1).

For a particular formation, if the magnetization of two (or more) limbs of folded rock strata are brought into agreement after correction for bedding, then the magnetization pre-dates the folding. MCELHINNY (1964) gives a statistical treatment of this test, and formations satisfying both criteria are said to have a positive fold test (age of magnetization pre-dates folding). The FISHER (1953) precision parameter k increases from 13.0 to 57.1 on restoring the Middle-Late Ordovician strata to the horizontal. This is significant at the 99% level according to MCELHINNY (1964). The age of magnetization is therefore pre-Benambran Orogeny, or Pre-Early-Middle Silurian in age. Both reversed and normal directions were present.

For the Silurian, eight out of a total of ten formations contained some remanence of the primary direction of magnetization. This characteristic primary direction was isolated somewhere in the temperature range 400°C–650°C (Fig. 3). Positive fold tests on the Silurian formations confirmed that the directions are primary. On bedding correction the resultant poles could be subdivided into two clusters. The oldest corresponds to an Early-Middle Silurian mean pole, the second to a Middle-Late Silurian mean pole (Fig. 4).

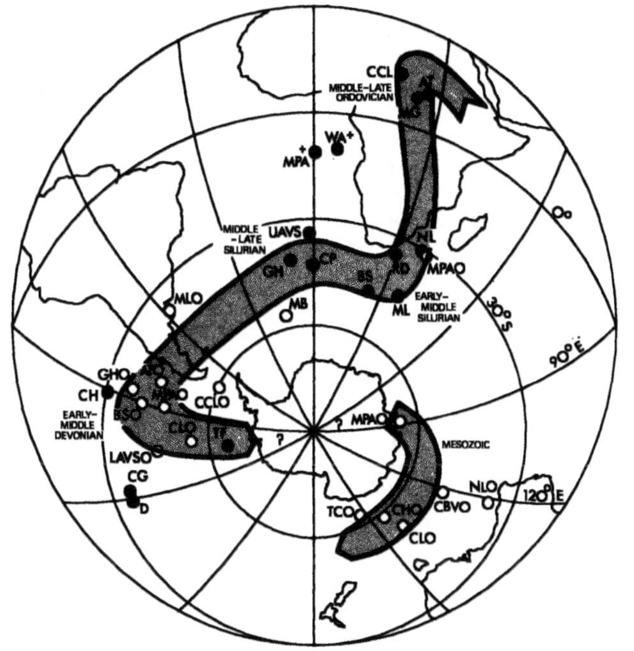

Fig. 4. Polar wander path constructed using new poles. The two poles marked with a cross are the only Early Ordovician poles obtained and complexities in the local structure may explain their position. The path is very well defined from the Ordovician to the Latest Silurian. The path between the Devonian and the Mesozoic may not be continuous, and for this reason is not connected. Solid circles, primary poles; Open circles, secondary poles. Mnemonic's used are listed in Table 2.

The mean Early-Middle Silurian direction is $D=45.9°$, $I=-8.3°$, with $k=37.8$ and $A_{95}=8.5°$ (Fig. 3). The mean Middle-Late Silurian direction is $D=19.1°$, $I=10.2°$, with $k=28.6$ and $A_{95}=9.2°$ (Table 1). Both reversed and normal directions were observed (Fig. 3).

The Devonian, although the youngest proved the most complex as this time interval corresponds to the period of major orogenic activity in the area. Thermal activity associated with folding resulted in a strong Devonian overprinting, with a direction co-linear with the primary direction. A primary direction for the Devonian is $D=320.9°$, $I=-24.4°$, with $k=11.2$ and $A_{95}=28.7°$ (Fig. 3; Table 1) was obtained. This compares with, and is supported by a strong Devonian mean secondary overprint direction obtained from both the Ordovician and Silurian Formations of $D=325.7°$, $I=-12.8°$, with $k=26.7$ and $A_{95}=9.5°$.

By eliminating the known present field and the primary direction, the remaining directions could be identified when compared with primary directions obtained from younger formations. In this way, the age of the secondary directions were obtained. For example, for the Ordovician Formations secondary directions isolated were, 1) a Silurian overprint (corresponding to minor folding during the Benambran Orogeny), 2) the strong Devonian overprint (major deformation during the Bowning Orogeny), 3) a poorly

Table 2. Palaeomagnetic data.

Formation	Mnemonic	N	Decl	Incl	k	A_{95}	Lat (°N)	Long (°E)	DP	DM
				Corrected primary directions						
Early Ordovician										
Walli And.	WA	25	225.59	−52.81	24.2	6.01	−10.46	6.26	5.73	8.30
Mt Pleasant And.	MPA	5	38.16	54.98	50.1	10.91	−12.20	0.01	10.98	15.48
Middle-Late Ordovician										
Cliefden Cvs Ls.	CCL	68	259.04	−58.58	28.6	3.28	13.20	20.21	3.61	4.89
Malongulli Fm.	MG	61	256.85	−52.18	63.7	2.30	8.07	24.68	2.16	3.15
Angullong Tuff	AT	32	259.65	−51.41	21.1	5.66	9.62	26.65	5.22	7.69
Early-Middle Silurian										
Rockdale Fm.	RD	12	45.17	1.41	63.9	5.47	−35.51	29.38	2.74	5.47
Millambri Fm.	ML	24	40.62	−18.38	21.3	6.57	−45.62	36.45	3.55	6.83
Belubula Shale	BS	19	33.47	−10.57	52.4	4.68	−48.00	23.96	2.40	4.74
Narragnal Ls.	NL	19	52.52	−1.43	174.8	2.55	−31.25	37.14	1.28	2.55
Middle-Late Silurian										
Canowindra Phy.	CP	9	21.46	15.71	20.7	11.57	−43.70	359.85	6.12	11.90
U. Avoca V. Sh.	UAVS	40	25.18	30.10	50.6	6.85	−34.74	358.64	4.22	7.60
Ghost Hill Fm.	GH	27	196.68	−22.77	37.2	4.62	−41.84	350.97	2.60	4.90
Mumbil Fm.	MB	78	188.69	1.18	10.5	5.20	−56.88	345.02	2.60	5.20
Early-Middle Devonian										
Dolerite intrus.	D	6	301.19	−34.02	103.6	6.61	−35.85	238.74	4.32	7.56
Cowra Granod.	CG	8	302.47	−31.62	30.2	10.24	−36.17	241.52	6.45	11.49
Cunningham Fm.	CH	17	312.38	8.10	52.9	4.95	−31.82	268.89	2.51	4.99
Tenandra Fm.	TF	11	154.78	41.87	26.7	9.00	−66.05	255.49	6.76	11.03
				Secondary Components						
Silurian overprint										
Mt Pleasant And.	MPAO	5	233.43	6.00	18.4	18.32	−31.63	39.43	9.24	18.40
Devonian overprint										
Angullong Tuff.	ATO	11	328.06	−1.48	40.8	7.24	−45.55	279.26	3.62	7.24
Mt Pleasant And.	MPAO	6	325.14	−13.22	47.8	9.79	−48.00	271.00	5.09	9.99
Millambri Fm.	MLO	13	157.40	−27.01	9.2	14.45	−37.54	301.77	8.55	15.72
L. Avoca V. Sh.	LAVSO	14	316.63	−27.00	41.8	6.22	−46.34	254.23	3.68	6.76
Canomodine Ls.	CLO	13	326.40	−32.26	27.8	8.01	−55.94	258.36	5.09	9.03
Cliefden Cvs Ls.	CCLO	19	341.60	−19.47	31.2	6.11	−60.97	289.08	3.33	6.38
Ghost Hill Fm.	GHO	6	319.54	2.64	174.2	5.09	−38.37	272.98	2.55	5.09
Belubula Shale	BSO	24	320.00	−5.50	20.3	6.73	−41.60	269.66	3.39	6.75
Mesozoic? overprint										
Canomodine Ls.	CLO	7	14.74	−79.19	12.1	18.07	−53.47	140.01	32.69	34.37
Mt Pleasant And.	MPAO	17	29.32	−60.91	19.5	8.29	−65.53	87.50	9.70	12.68
Cugga Burga Vol.	CBVO	21	41.46	−75.75	32.0	5.70	−50.13	121.12	9.65	10.49
Narragnal Ls.	NLO	15	70.70	−77.80	18.6	9.12	−37.31	120.89	16.09	17.13
Tolga Calcaren.	TCO	46	357.09	−73.86	86.0	2.28	−63.62	152.06	3.70	4.11
Cunningham Fm.	CHO	18	7.27	−75.90	13.4	9.81	−59.06	142.68	16.65	18.08

defined Mesozoic overprint (Mesozoic activity associated with the break up of Australia and Antarctica during the Createous), and 4) the present field direction.

5. Discussion

The primary pole positions obtained are shown in Table 2 and these indicate a well defined apparent polar wander path for the Early Palaeozoic of South East Australia (Fig. 4). These new poles, when plotted with those used in the previous interpretations produce a stratigraphically coherent path (Fig. 5). The general trend of the Central Australian apparent polar wander path of McElhinny and Embleton (1974), defined by the Cambro-Ordovician pole positions LFG, JF, and DG to the Ordovician TS, AD, and the Siluro-Devonian MS pole position is the same as the trend defined by the new data (Fig. 5).

The new data for the Middle-Late Silurian however, indicate the start of a sharp leftwards swing of the path towards South America. The pole path then swings back

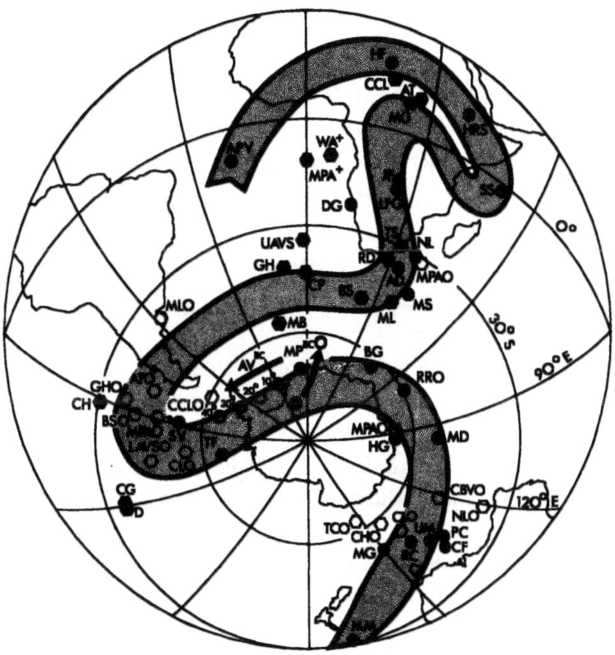

Fig. 5. Combining the mean of new poles for each time period and the other Australian poles, a single coherent path for Australia since the Late Precambrian is constructed. With this path three poles are in error, SV, AV, and MP. On bedding correction, the MP pole moves towards a stratigraphically correct position on the new path. The AV pole, however, moves towards the Devonian region of the path. This suggests both the SV and AV poles are secondary Devonian poles. See text for description. Solid circles are Australian poles listed in Fig. 1. Solid hexagons are new primary poles; open hexagons are new secondary poles (Mnemonic's are listed in Table 2). Open circles represent the bedding corrected Ainslie Volcanics (0–40° dip correction) and the Mugga Porphyry.

during the Early-Middle Devonian? (Fig. 5), corresponding with the path previously identified as representing the separate plate motions of South East Australia (MCELHINNY and EMBLETON, 1974).

If the ages of magnetization for the new Devonian pole are correct then the poles SV, MP, and AV plot out of sequence. Recent work in the Canberra area (M. Owen, personal communication) has shown that the Ainslie Volcanics (AV) and the Mugga Porphyry (MP) both previously thought to be flat lying, in fact have substantial dips. The Mugga Porphyry (MP) has an average south-westerly dip of 27° whereas the Ainslie Volcanics dip to the north-east at angles up to a maximum of 35°. Stratigraphic relations indicate the Ainslie Volcanics (old pole pos. 71°S, 353°E) are Late Wenlockian and not Early Devonian as previously documented, and therefore older than the earliest Ludlovian Mugga Porphyry (M. Owen, personal communication), (old pole pos. 80°S, 340°E). When these bedding corrections are applied, the Mugga Porphyry moves to the Middle-Late Silurian region (new pole pos. 63.2°S, 007.9°E; Table 1) of the new path (Fig. 5), but the Ainslie Volcanics move towards the Devonian region. The precise dip of the Ainslie Volcanics at the sampling sites is unknown, however dips range from horizontal to 35° about a strike of 179°, within the area sampled. Bedding corrected poles for a constant strike and varing dip range from 72.3°S, 341.2°E for 5° dip to 67.7°S, 285.3°E for a dip of 35° (Table 1). This suggests that the Ainslie Volcanics as measured by LUCK (1973) contain this strong Devonian overprint. The main folding occurred after the Devonian overprinting, during the Upper Devonian-Lower Carboniferous.

The Silurian Volcanics (SV) pole position lies on the new Devonian direction, suggesting this direction is in fact indicating the strong Devonian overprint direction found throughout the Ordovician and Silurian. Neither adequate cleaning nor has any fold test been applied on these samples. Further work on both the Ainslie Volcanics and the Silurian Volcanics using present techniques will prove or disprove this suggested explanation of the SV discrepency.

Small scale rotations between the tectonic provinces sampled and neighbouring tectonic provinces or the stable central Australian craton were not detectable, though geological evidence indicates small rotations did occur (SCHEIBNER, 1974). The 20° discrepency between the new South East Australian Ordovician pole and the central Australian Ordovician pole may indicate such tectonic movement.

I am grateful to Drs. M. W. McElhinny and C. T. Klootwijk for helpful discussion during the various stages of the work and to Mr. D. J. Edwards who assisted greatly in the operation and maintenance of the instruments. All are members of the Australian National University. My thanks also to Drs. M. Owen and G. A. M. Henderson of the Bureau of Mineral Resourses, Canberra, Australia, for valuable discussion concerning the age, structure, and geology of the Ainslie Volcanics and the Mugga Porphyry. Critical reading of the manuscript by M. W. McElhinny, C. T. Klootwijk, B. J. J. Embleton, and P. W. Schmidt was greatly appreciated.

REFERENCES

BUREK, P. J., An advanced device for chemical demagnetization of red beds, *Z. Geophys.*, **37**, 493–498, 1971.

EMBLETON, B. J. J., The palaeomagnetism of some Palaeozoic sediments from central Australia, *J. Proc. R. Soc. N. S. W.*, **105**, 86–93, 1972.

EMBLETON, B. J. J. and D. J. EDWARDS, An instrument for orienting rock samples, *J. Phys. Earth Sci. Inter.*, **6**, 672, 1973.

EMBLETON, B. J. J. and J. W. GIDDINGS, Late Precambrian and Early Palaeozoic palaeomagnetic results from South Australia and Western Australia, *Earth Planet. Sci. Lett.*, **22**, 355–365, 1974.

EMBLETON, B. J. J., M. W. MCELHINNY, A. R. CRAWFORD, and G. R. LUCK, Palaeomagnetism and the tectonic evolution of the Tasman Orogenic Zone, *J. Geol. Soc. Aust.*, **21**, 187–194, 1974.

FISHER, R. A., Dispersion on a sphere, *Proc. R. Soc. Lond.*, **A217**, 295–305, 1953.

GOREE, W. S. and M. FULLER, Magnetometers using RF-Driven SQUIDS and their applications in Rock Magnetism and Palaeomagnetism, *Rev. Geophys. Space Phys.*, **14**, 591–608, 1976.

GRAHAM, J. W., The stability and significance of magnetism in sedimentary rocks, *J. Geophys. Res.*, **54**, 131–167, 1949.

LUCK, G. R., Palaeomagnetic results from Palaeozoic sediments of northern Australia, *Geophys. J.*, **28**, 475–487, 1972.

LUCK, G. R., Palaeomagnetic results from Palaeozoic rocks of southeast Australia, *Geophys. J.*, **32**, 35–52, 1973.

MCELHINNY, M. W., Statistical significance of the fold test in palaeomagnetism, *Geophys. J. R. Astr. Soc.*, **8**, 338–340, 1964.

MCELHINNY, M. W., An improved method for demagnetizing rocks in alternating magnetic fields, *Geophys. J. R. Astr. Soc.*, **10**, 369–374, 1966.

MCELHINNY, M. W. and B. J. J. EMBLETON, Australian palaeomagnetism and the Phanerozoic plate tectonics of eastern Gondwanaland, *Tectonophysics*, **22**, 1–29, 1974.

MCELHINNY, M. W., G. R. LUCK, and D. J. EDWARDS, Large volume magnetic field-free space for thermal demagnetization and other studies in palaeomagnetism, *Pure Appl. Geophys.*, **90**, 126–130, 1971.

MOLYNEUX, L., A complete result magnetometer for measuring remanent magnetization of rocks, *Geophys. J.*, **24**, 429–433, 1971.

MOREL, P. and E. IRVING, Tentative palaeocontinent maps for the Early Phanerozoic and Proterozoic, *J. Geol.*, **86**, 535–561, 1978.

PACKHAM, G. H. (editor), The geology of New South Wales, *J. Geol. Soc. Aust.*, **16**, 1–654, 1969.

SCHEIBNER, E., A plate tectonic model of the Palaeozoic history of New South Wales, *J. Geol. Soc. Aust.*, **20**, 405–426, 1973.

SCHMIDT, P. W. and W. A. MORRIS, An alternative view of the Gondwana Palaeozoic apparent polar wander path, *Can. J. Earth Sci.*, **14**, 2674–2678, 1977.

WELLS, A. T., D. T. FORMAN, L. C. RANFORD, and P.J. COOK, Geology of the Amadeus Basin, central Australia, *Bull. Bur. Miner. Res. Geol. Geophys. Aust.*, 100, 1970.

ZIJDERVELD, J. D. A., A. C. Demagnetization of rocks, in *Methods in Palaeomagnetism*, edited by D. W. Collinson, K. M. Creer, and S. K. Runcorn, pp. 254–286, Elsevier, New York, 1967.

Palaeomagnetism and Reconstruction of Palaeogeographic Positions of the Siberian and Russian Plates during the Late Proterozoic and Palaeozoic

A. N. Khramov and V. P. Rodionov

All-Union Oil Research Geological Prospecting Institute, Leningrad, USSR

(Received December 13, 1979 ; Revised May 14, 1980)

An analysis of the palaeomagnetic data from the Siberian and Russian plates, Kazakhstan, and the Scythian and Turanian blocks permits us to reconstruct the history of their horizontal movements from the Late Proterozoic through to the Triassic. The Siberian plate originally consisted of four separate blocks: the Anabar craton, the Aldan shield, and the northern and western Baikalian blocks. Situated in southern subequatorial latitudes during the late Proterozoic, these separate units collided at the beginning of the Palaeozoic to form the Siberian plate. During the Palaeozoic the Siberian plate moved northward along a great circle path through the subtropics, across the equator, and eventually past the North Pole. During the early and middle Palaeozoic the Russian plate was located in low southern latitudes. It moved northward during the Palaeozoic, colliding with Siberia, Kazakhstan, and the Scythian and Turanian blocks during the Permian.

1. Introduction

The history of the late Proterozoic and Palaeozoic plate motions which gave rise to the present structure of most of northern Eurasia is an important, and yet unsolved, problem of global tectonics. Several models have been proposed, especially for pre-Devonian times, however, much ambiguity and uncertainty remains (Briden *et al.*, 1974; Gorodnitsky *et al.*, 1978; Morel and Irving, 1978; Bucha, 1975). A key problem in reconstructing the positions of the continents during the late Proterozoic and Palaeozoic is the relative positions of the Siberian and Russian plates.

The use of palaeomagnetic data to unravel the history of the Palaeozoic movements of the Russian and Siberian plates began as far back as the 60's when the first set of palaeomagnetic determinations for the Palaeozoic rocks of the Russian and Siberian platforms were published (Forsch and Khramov, 1961; Khramov *et al.*, 1965). These data indicated that during the early Palaeozoic both platforms not only moved with respect to the geographical pole but also moved with respect to each other (Khramov, 1965). An alternate hypothesis was later put forward which explained that the divergence of early Palaeozoic poles was the result of late Palaeozoic remagnetization (Creer, 1970). However, it now seems certain that there has been considerable relative motion between the Russian and Siberian plates (Khramov *et al.*, 1974).

2. Palaeomagnetic Poles and the Trajectories of Their Apparent Movement

Catalogs of palaeomagnetic data from the USSR (KHRAMOV, 1971, 1973, 1975, 1979) contain 165 determinations for the Palaeozoic and early Triassic of the epi-Baikal platforms, and 105 determinations from the Ural mountain belt, Kazakhstan block, and Turanian and Scythian plates. These determinations were analyzed and selected on the basis of the following criteria:

a) The geological age of the units are known to an accuracy of epoch and there is no evidence that the magnetization is not contemporaneous.

b) The time interval sampled is sufficient to account for secular variation.

c) The number of independently oriented samples is not less than ten; and the A_{95} confidence interval does not exceed 15 degrees.

d) A primary (characteristic) direction of mangetization is recognized.

e) Local tectonics have been taken into account.

Several palaeomagnetic determinations from the Baltic shield which belonged to confidence group "A" of Irving were added to the East European group (IRVING *et al.*, 1976; MCELHINNY and COWLEY, 1977). The palaeomagnetic determinations chosen for the Siberian plate, Altai-Sayan region, Russian plate, Urals (A and B), Kazahkstan block, and the Turanian and Scythian plates have been grouped into time intervals of one to two geological epochs in length. A mean pole was calculated for each group. The results are shown in Table 1 and Fig. 1.

The palaeomagnetic poles for the early Palaeozoic of the Siberian platform cluster tightly ($A_{95}=3°–8°$). From the early Cambrian through the Ordovician the pole migrates westward from a position southwest of New Zealand, across Australia, towards the equator. Middle Palaeozoic poles from the Siberian plate are not as numerous and do not group as tightly ($A_{95}=12°–24°$). Unfossiliferous redbeds and undated intrusives were used to determine the Middle Palaeozoic section of the apparent polar wander path; therefore, the poor clustering may be the result of erroneous age assignments. The Permian and Triassic determinations from the Siberian plate are numerous ($N=27$) and group very well ($A_{95}=4°–8°$). During the Permo-Triassic the North Pole was situated in the Sea of Okhotsk, east of Sakhalin.

Though there are not many determinations from the Altai-Sayan region, those determinations which have been made diverge significantly from the Siberian data, and therefore, should be treated separately. The Altai-Sayan poles are displaced to the east of the Siberian polar wander path during the early Palaeozoic, and to the south of the polar wander path during the middle Palaeozoic (Table 1, Fig. 1a).

For the Russian platform there are numerous palaeomagnetic poles from the late Palaeozoic and early Triassic ($N=52$), and they cluster very tightly (Table 1). Though early and middle Palaeozoic palaeomagnetic poles also cluster well, there is a possibility that they may have been remagnetized during the late Palaeozoic.

Several early Carboniferous and late Devonian determinations from the Russian platform give anomalous directions. One determination (not shown in Table 1) from Tournaisian redbeds of the northern Moscow basin gave a pole at, $\phi=7$, $\Lambda=133$ (KHRAMOV *et al.*, 1974, p. 136). A similar pole was obtained by thermal cleaning of some

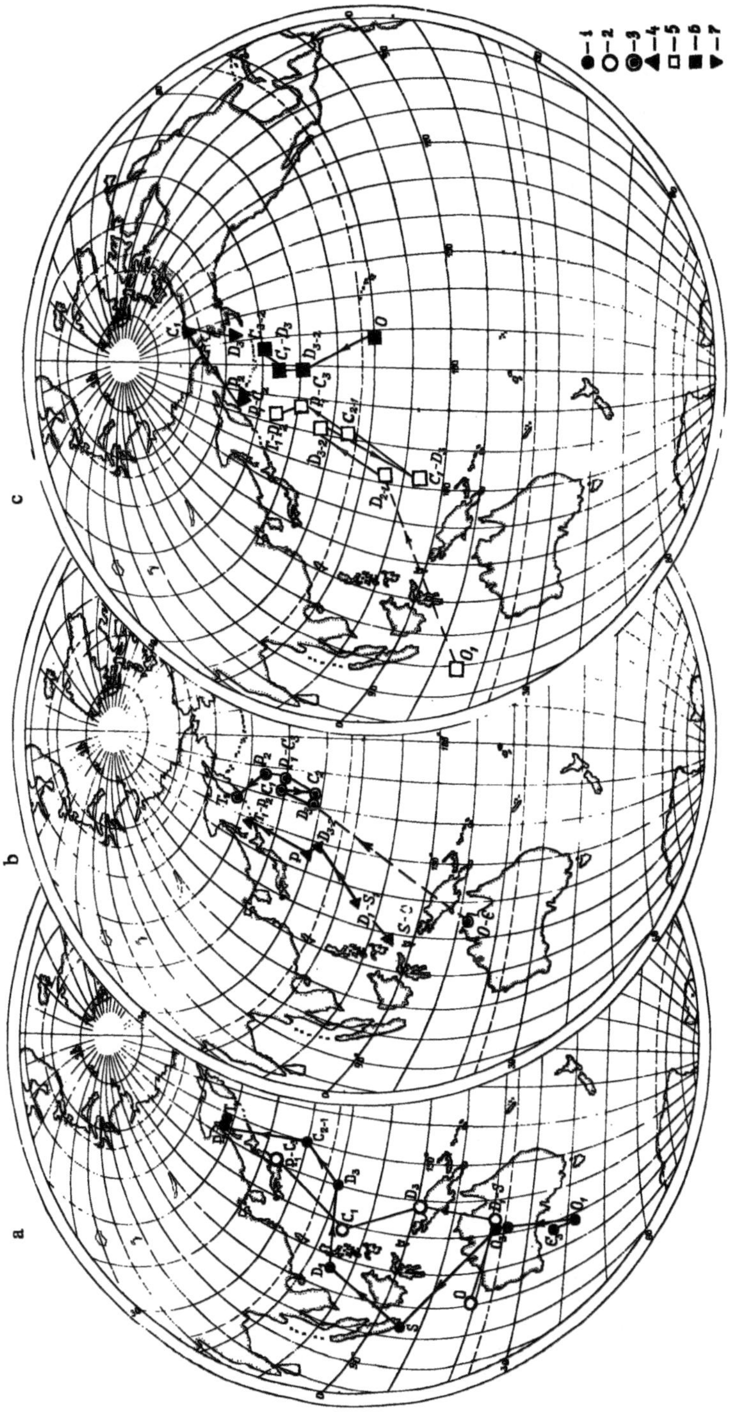

Fig. 1. Trajectories of the apparent palaeomagnetic pole wandering: a, for the Siberian plate (1) and Altaian-Sayanian region (2); b, for the Russian platform (3) and Turanian and Scyphian plates (4); c, for the Urals and Kazakhstan; (5), Urals, group A; (6), Urals, group B; (7), Kazakhstanian block.

Table 1. The mean palaeo-

Geological epochs	Catalogue numbers (KHRAMOV, 1971, 1973, 1975, 1979; MCELHINNY, 1972, 1977)	The number of data	The mean palaeomagnetic pole		
			Φ, N	Λ, E	A_{95}°
1	2	3	4	5	6
	I. Siberian plate				
T_1	6-19, 6-20, 6-21, 6-22, 6-27, 6-28, 6-33, 6-34, 6-37, 6-38, 6-39, 6-47, 6-48, 6-50, 6-52, 6-53, 6-55, 6-56, 6-68, 6-69	20	52	147	4
P_2	6-49, 6-60, 6-61, 6-62, 6-70, 6-71, 7-40	7	52	145	8
C_2+C_1	8-36, 8-37, 8-38, 8-45	4	30	150	14
D_3	9-19, 9-43, 208, 213	4	20	140	12
D_1	9-32, 9-34, 9-36	3	15	116	24
S_1	10-07, 238	2	−7	103	—
O_3	11-02, 11-03, 11-12, 11-13, 11-14, 11-23	6	−21	131	6
O_2	11-04, 11-05, 11-06, 11-07, 11-08, 11-15	7	−24	132	4
O_1	11-16, 11-17, 11-18, 11-19, 11-20	5	−42	127	6
$€_3$	12-03, 12-04, 12-05, 12-06, 12-07, 12-08, 12-17, 12-18, 12-19, 12-21, 12-22, 12-23, 12-26, 12-27, 12-41, 12-42, 12-43	17	−36	126	3
$€_2$	12-13, 12-28, 12-29, 12-30	4	−44	156	5
$€_1$	12-24, 12-32, 12-33	3	−44	157	8
	II. Altajan-Sayanian region				
$P+C_3$	7-41, 7-42, 7-45, 8-34	4	37	141	13
C_1	8-46, 8-47, 8-49, 9-50	4	16	128	12
D_3	8-48, 9-26, 9-40, 9-44	4	−1	138	23
D_1+S	9-41, 9-45, 9-46, 9-47, 10-11	5	−21	134	9
O	12-02	1	−19	110	—
$€_1$	12-31	1	−45	53	—
	III. East-European (Russian) plate				
T_1	6-08, 6-09, 6-13, 6-14, 6-16, 6-30, 6-43, 6-44, 6-45, 6-46, 6-65	11	53	156	5
P_2	7-01, 7-02, 7-06, 7-13, 7-14, 7-15, 7-16, 7-17, 7-23, 7-24, 7-25, 7-26, 7-27, 7-28, 7-29, 7-31, 7-32, 7-33, 7-35, 7-36, 7-48, 7-49	23	46	167	2, 5
P_1+C_3	7-19, 7-20, 7-21, 7-55, 7-56, 8-01, 8-02, 8-03, 8-55, 8-56, 6.94, 6.126, 6.127, 6.324, 7.113, 14/316, 14/336, 14/341	18	40	166	6

magnetic poles for the palaeozoic.

Geological epochs	Catalogue numbers (KHRAMOV, 1971, 1973, 1975, 1979; MCELHINNY, 1972, 1977)	The number of data	The mean palaeomagnetic pole		
			Φ, N	Λ, E	A_{95}°
1	2	3	4	5	6
C_2	8-05, 8-06, 8-07, 8-09, 8-51, 8-52, 6.104, 14/353	9	32	163	4
C_1	8-12, 8-13, 8-14, 8-15, 196	5	41	163	8
D_3	9-01, 9-02, 9-03, 9-04, 9-05	5	32	161	4
D_1+S_2	4-30, 5-75, 9-16, 10-08, 10-09	5	35	160	11
O	11-09, 11-10, 14/107, 4*1, 5*2	5	30	143	9
O−C	12-01, 2*1	2	−13	135	—
	IV. Turanian and Schythian plates				
T_1+P_2	6-17, 7-09, 7-47	3	48	149	13
P_1	7-54, (2-3)*3, (4-6)*4	6	31	146	6
D_3+D_2	9-37, 9-50, 9-51, 9-52	4	29	149	20
D_1+S	9-55, 10-15	2	16	136	—
S−O	10-05	1	6	128	—
	V. Urals, group A				
T_1+P_2	6-05, 6-06, 6-07, 6-66, 7-88, 7-39, 7-50	7	46	163	11
P_1+C_3	8-32, 8-33, 8-55, 8-56, 8-57	6	40	168	6
C_2+C_1	8-24, 8-29, 8-39, 8-41, 8-42, 8-58, 8-64	7	27	160	11
C_1-D_3	8-18, 8-28, 9-27, 9-39	4	7	151	4
D_3-D_2	9-07, 9-17, 9-18, 9-21, 9-23, 9-29, 9-38, 9-48, 9-49, 8-43, 8-65	11	33	163	8
D_1+S_2	9-15, 9-57, 10-12, 10-14	4	15	151	19
O_1	11-22	1	−14	96	—
	VI. Urals, group B				
C_3-C_2	8-17, 8-23, 8-31, 8-63	4	49	186	12
C_1-D_3	8-40, 8-44, 8-65, 9-06	4	46	177	6
D_3+D_2	9-07, 9-08, 9-23, 9-30	4	39	179	11
S_1	10-02, 10-03, 10-16, 10-18, 10-19	5	9	166	17
O	10-17, 11-01, 11-21, 12-34	4	20	188	8
	VII. Kazakhstanian block				
P_2	13/43	1	54	169	—
P_1-C_2	7-22, 8-04, 179, 180	4	54	167	10
C_1	13/60, 13/61, 13/62	3	69	201	4
D_3	202	1	56	193	—

Φ, Λ, Latitude and longitude of the palaeomagnetic pole; A_{95}°, radius of the circle of confidence for the mean pole.

*1 KHRAMOV *et al.*, 1974; *2 KHRAMOV and RODINOV, 1977; *3 ASANIDZE and PECHERSKY, 1979; *4 Ju. V. Kariakin and A. N. Khramov, new data on the North Caucasian red beds.

late Devonian redbeds from Valdai: $\phi=-8$, $\Lambda=133$ (KHRAMOV *et al.*, 1974, p. 137). Detailed studies of the Ordovician of the Baltic area also give an anomalous direction corresponding to a pole at $\phi=13, \Lambda=133$ (BUCHA *et al.*, 1976; KHRAMOV and RODIONOV, 1977).

These poles are similar to early Carboniferous and late Devonian poles from Group A of the Urals (Table 1). A single pole from the early Ordovician of the Urals (A) lies at similar latitudes, but farther to the west. The remaining poles of Group A for the Urals coincide with coeval poles from the Russian platform, indicating that they were part of the same plate. Most of the determinations assigned to group A were obtained from rocks west of the so-called "Major Fault" of the Urals, and hence, were tectonically situated within the Russian plate. In contrast, all the determinations from Group B of the Urals were taken from the greenstone belt associated with the "Major Fault." Geologic and palaeomagnetic data suggest that thrusting, sinistral faulting, and counterclockwise rotation of the rock units in this area have taken place (IVANOV, 1968). This rotation probably is responsible for the eastward shift of the polar wander curve of Group B (Fig. 1c).

Similarly, the palaeomagnetic determinations for the Turanian and Scythian plates are from rocks of severely deformed regions. The southwest shift of poles from these areas with respect to contemporaneous poles of the Russian platform (Table 1) may be due to clockwise rotation along dextral faults.

Palaeomagnetic results from the Kazakhstan block are displaced far to the northeast of all other determinations. From the Devonian through to the end of the Palaeozoic the apparent polar wander curve for Kazakhstan lies to the east of the curve for Group B of the Urals (Fig. 1c).

3. Interpretations of the Apparent Polar Wander Curves and Plate Motion

An analysis of the palaeomagnetic data listed in Table 1 enables us to construct the apparent polar wander paths for the Siberian and Russian plates during the Palaeozoic.

Siberian plate. A smooth apparent polar wander path can be drawn through the palaeomagnetic determinations listed in Table 1. The data are best known for the Cambro-Ordovician and Permo-Triassic; the middle Palaeozoic part of the curve should be considered approximate. Though there are only a few Middle Palaeozoic determinations, the inferred position of the polar wander path is confirmed by palaeomagnetic determinations from the Altai-Sayan region. The poles from this area form a similar, though southward shifted, apparent polar wander curve (Table 1).

Russian plate. The early Palaeozoic part of the polar wander path for the Russian plate is based on a single late Cambrian/early Ordovician pole, $\phi=-13$, $\Lambda=135$. Ordovician data from the Urals support this position if the directions are rotated back to their presumed original orientation. The early Palaeozoic poles from Ural groups A and B are nearly coincident if they are rotated clockwise 35 degrees, and counterclockwise 55 degrees, respectively.

A pole at $\phi=8, \Lambda=147$, which is the average of pole positions IV S$-$O and VI S_1 (Table 1) is proposed for the early Silurian. This averaged pole position allows for the rotations which may have taken place within the eastern Urals and Scythian block.

The Middle palaeozoic section of the polar wander curve is composed of poles V D_1+S_2, III D_3, V C_1-D_3, and III C_1. Poles III O and III D_1+S_2 have not been included because we believe that they have been remagnetized and correspond to a pole position of late Paleozoic age. The widespread effects of late Palaeozoic remagnetization have recently been reported from the Russian plate, the Urals, and within the Hercynides of Central Europe (KHRAMOV *et al.*, 1974; BUCHA *et al.*, 1976).

Hence, the major difficulty facing a study of the horizontal movement of the Russian and Siberian plates is the unreliability of information from the middle Palaeozoic of both of these areas, and the scarcity of palaeomagnetic data from the early Palaeozoic of the Russian platform, proper. Nevertheless, the inferred polar wander paths permit us to model the horizontal movements of both plates and to reconstruct their relative positions during the Palaeozoic and early Triassic.

The following principles were used in making the reconstructions. First, the ancient magnetic field was assumed to be a geocentric dipole. Reconstructions were made by rotating all the palaeomagnetic poles so that they were coincident with the geographic pole. Secondly, the problem of longitudinal uncertainty was dealt with by minimizing the longitudinal separation between both plates.

The reconstructions shown in Figs. 2 and 3 allow us to trace the major features of the relative movements of the Russian and Siberian plates. At the beginning of the Cambrian, the Siberian plate was situated in the southern hemisphere with its present southern edge oriented to the north (Fig. 3). From the Cambrian through the end of the Ordovician the Siberian plate moved northward from a latitude of 10°–30° south to a belt 5°S–15°N, i.e., travelling a distance of 25 degrees (approximately 3,000 km). During the late Palaeozoic the present day eastern edge of Siberia (Verkhoyansk mountains) approached the North Pole, passing near the pole in Permo-Triassic time (Fig. 3).

Similarly, during the early Palaeozoic the Russian plate was situated in the southern hemisphere at low latitudes. At the close of the Silurian the equator ran approximately parallel to the Urals. During the Tournasian and into the middle Carboniferous the Russian plate moved in a retrograde fashion, shifting toward the equator. The times of most rapid plate motion were at the close of the Ordovician/beginning of the Silurian, and during the Middle Devonian, early Carboniferous and late Permian/Middle Triassic.

From the late Silurian through the early Carboniferous, the Altai-Sayan foldbelt underwent a clockwise rotation with respect to the Siberian platform. However, in the Urals this motion was accompanied by a counterclockwise rotation which commenced in the Silurian and probably did not end until the early Permian.

The Turanian and Scythian plates moved together with the Russian platform during the middle and late Palaeozoic. The divergence of the polar wander paths is apparently the result of clockwise rotations along dextral faults. The Kazakhstan block moved slowly during the middle and late Palaeozoic; the early Palaeozoic history of its movements is unknown. Table 1 shows that by the end of the Palaeozoic, Kazakhstan, the Siberian and Russian plates, along with their associated fold belts, were welded together to form the nucleus of northern Eurasia.

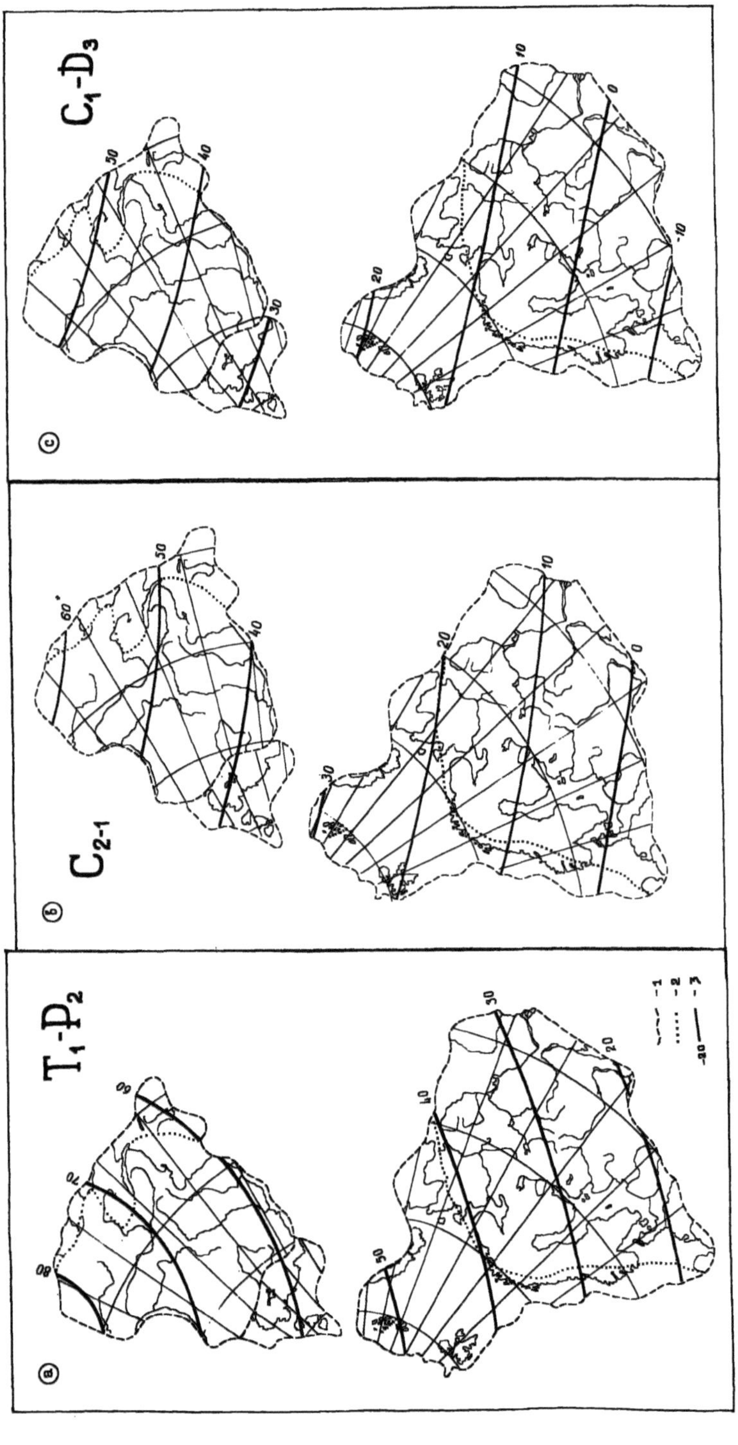

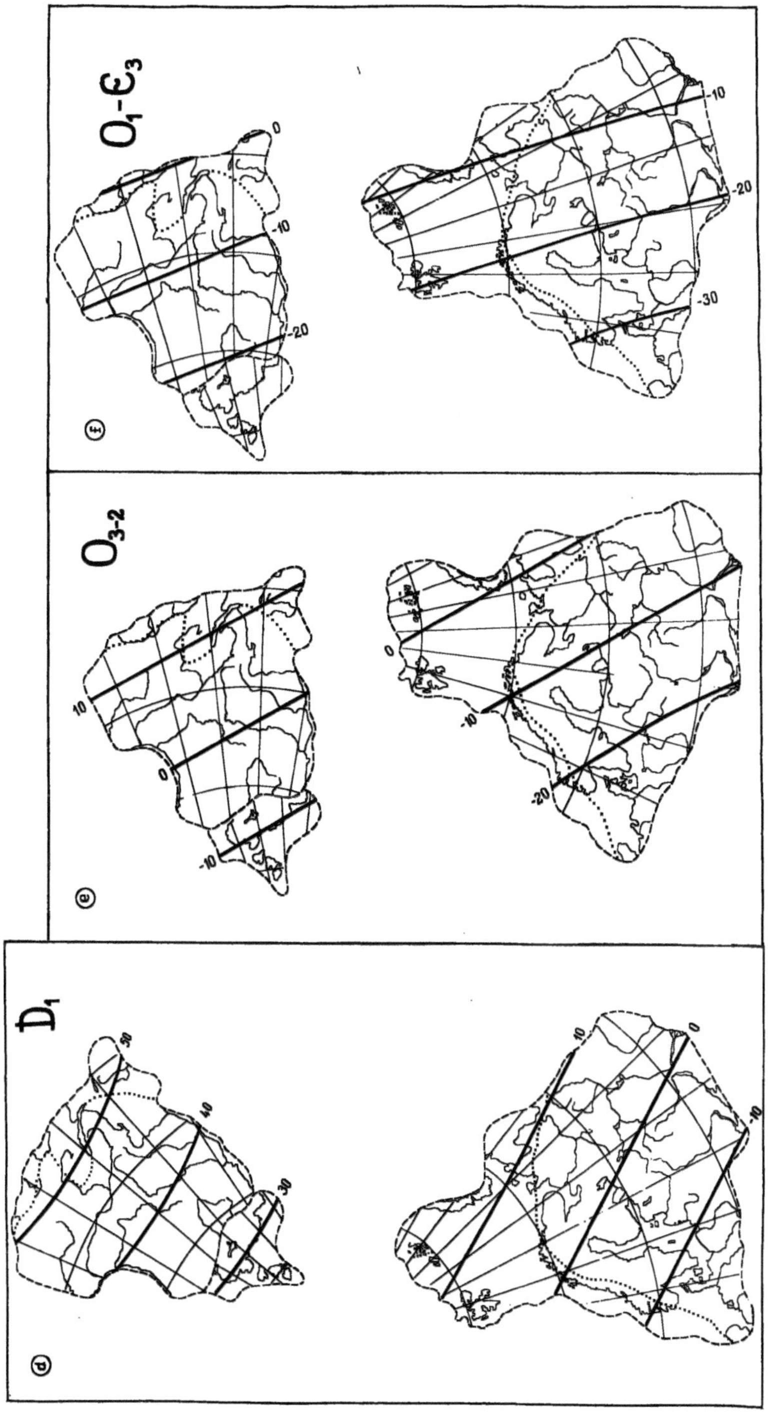

Fig. 2. Reconstructions of the positions of Siberian and Russian platforms for the different epochs. a) Early Triassic and Late Permian, b) Middle and Early Carboniferous-Late Devonian, d) Early Devonian, e) Late and Middle Ordovician, f) Early Ordovician and Late Cambrian. 1, the epihercynian platform boundary; 2, the precambrian platform boundary; 3, palaeomagnetic latitudinal lines.

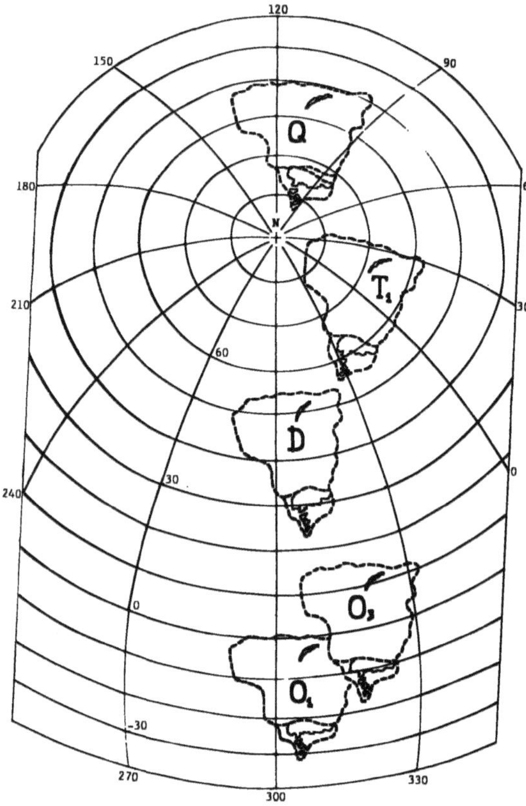

Fig. 3. Siberian plate drift since Early Palaeozoic.

4. Continental Reconstructions for the Late Proterozoic and Early Palaeozoic

Figure 2 illustrates the relative positions of the Siberian and Russian plates; however, due to longitudinal uncertainty these positions can not be considered to be unique. Alternate reconstructions of the early Palaeozoic have also been proposed (BRIDEN *et al.*, 1974; GORODNITSKY *et al.*, 1978; MOREL and IRVING, 1978; BUCHA, 1975). Briden's reconstruction differs from the reconstruction shown in Fig. 2 by inverting the longitudinal order of the plates. In this alternate reconstruction the Verkhoyansk edge of the Siberian plate is placed opposite the Urals (BRIDEN *et al.*, 1974; MOREL and IRVING, 1978). The reconstructions of BRIDEN *et al.* (1974) and MOREL and IRVING (1978) suggest that during the Ordovician and Silurian a very wide ocean (12.0×10^3 km) separated the Siberian and Russian plates. If this were the case, the resulting rates of plate convergence would have had to have been higher than those recorded during the rapid northward drift of India during the Cenozoic. Such a reconstruction is difficult to reconcile with Precambrian magnetic anomaly patterns, geological structures and metallogenic zones (GUSEV, 1973; TAMRAZYAN, 1971). These data suggest that in the late Proterozoic, the Russian and Siberian platforms, together with Kazakhstan formed a single landmass which differed from the present configuration by the nearly complete closure of the west Siberian lowlands.

In contrast, the reconstructions of Bucha (1975) and Gorodnitsky *et al.* (1978) orient the Russian and Siberian plates by reversing the polarity of Russian or Siberian early Palaeozoic poles. This is a more natural solution becuase it does not require such major displacements or rapid movements, and because it provides for the transition between late Proterozoic and early Palaeozoic configurations. However, even this solution is not free from problems. Questions still need to be answered regarding the complex nature of the middle Palaeozoic polar wander curve. In addition, an analysis of early Palaeozoic polarity profiles from each of these plates reveals significant discrepancies which cast doubt on the validity of these models.

Thus, the relative positions of the Russian and Siberian plates shown in Fig. 2 require further study. Though adequate palaeomagnetic data exists for Siberia (Khramov *et al.*, 1965, 1974; Khramov and Sholpo, 1967) and partly for Australia (Kirschvink, 1978), a final choice among the three reconstructions will not be possible until more complete palaeomagnetic profiles from the Lower Palaeozoic of Laurasia and Gondwana are assembled.

Palaeomagnetic data concerning the Precambrian history of the Russian and Siberian platforms are scarce and difficult to interpret. However, some preliminary data from Siberia may be presented. Recently the Vendian and Riphean sections were studied by the palaeomagnetic laboratory of VNIGRI. The results of this investigation

Table 2

Geological epochs	Catalogue numbers (Khramov, 1971, 1973, 1975, 1979)	The number of data	The mean palaeomagnetic pole		
			Φ, N	Λ, E	A_{95}°
	1. Anabarian shield with surrounding areas				
O_1	11-19	1	−42	134	—
$€_3$	12-21	1	−35	130	—
$€_2$	12-32	1	−39	153	—
	13-16	1	−14	144	—
PR_{3-2}	13-04	1	−20	108	—
	2. Aldanian shield				
$€_2$	12-11, 12-12, 12-13	3	−47	163	15
$€_1$	12-16	1	−54	208	—
R_{3-2}	13-05	1	−24	208	—
	3. North-Baikalian block				
V	12-24, 13-14, (3-5)*	5	−41	138	18
R_3	(1-4)*	4	−40	136	25
R_2	(1-3)*	3	−35	155	18
R_1	13-15, 2*	2	−26	90	
	4. Western Pre-Baikalian area				
€	1*	1	−30	148	—
V	1*	1	−36	165	—
R_3	1*	1	−31	159	—
R_2	(1, 2)*	2	−39	203	—

* New data by E. L. Gurevich, R. A. Komissarova, E. P. Osipova, S. A. Pisarevsky and V. P. Rodionov.

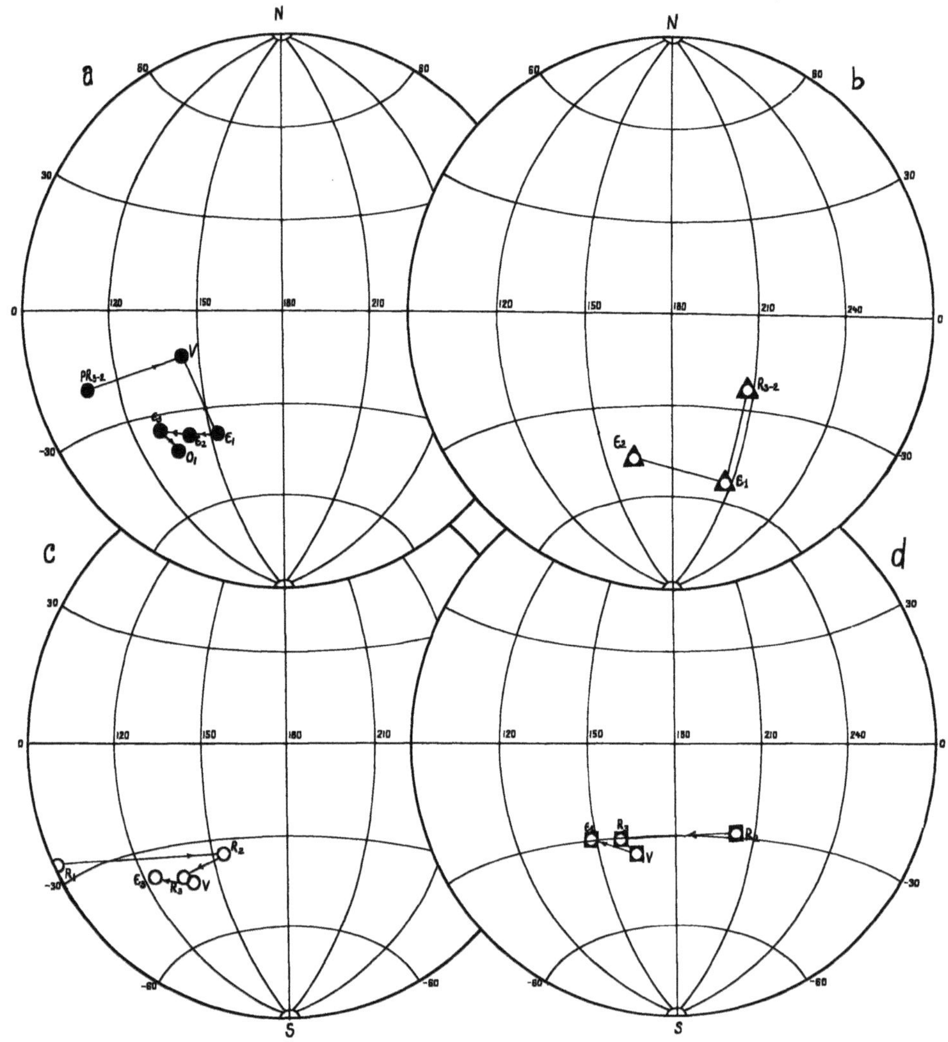

Fig. 4. Trajectories of the apparent palaeomagnetic pole wandering in Proterozoic times for Siberian plate blocks. a, Anabarian shield with surrounding areas; b, Aldanian shield; c, North-Baikalian blcok; d, West-Baikalian block.

supplement the data which has already been published (KHRAMOV, 1971, 1973, 1975; IRVING *et al.*, 1976; MCELHINNY and COWLEY, 1977).

The Vendian and Riphean non-metamorphosed deposits from the North and South Baikal area were studied and additional palaeomagnetic data were obtained from the Altai-Sayan region. Zones of normal and reverse polarity are present within the Vendian/Riphean sections, and these zones can be correlated laterally. All this evidence suggests a stable remanent magnetization.

The distribution of NRM directions indicates that though the directions are consistent within each area, they differ from region to region. The directions begin to group consistently only after the Late Cambrian (Table 2 and Fig. 4).

New palaeomagnetic data for the Proterozoic of the Siberian plate suggest that during the late Precambrian, Siberia was not a single block. Rather, there were four separate blocks: the Anabar massif and adjacent areas, the Aldan shield, and the north and west Baikalian blocks. During the Proterozoic all these blocks were located in southern, subequatorial latitudes, and more importantly were oriented differently with respect to one another than they are at present. By the Late Cambrian these blocks had collided, suturing to form the Siberian plate (Baikalian orogeny). This interpretation does not require the great amount of apparent polar wander suggested by McElhinny (1973, p. 217). If each block is treated separately, the amount of polar wander required is minimized.

5. Conclusion

Palaeomagnetic data for the Siberian and Russian plates enables us to reconstruct the history of their horizontal movements since the end of the Proterozoic. During the late Proterozoic, Siberia was divided into four separate blocks, each situated at low latitudes in the southern hemisphere. At the start of the Palaeozoic these blocks were brought together, and during the Baikalian orogeny (Late Cambrian) sutured to form the Siberian plate. During the Cambrian and Ordovician the Siberian plate, oriented with its present southern margin facing north, moved northward across the south subtropics towards the equator. During this interval the plate moved from latitudes 30°–10°S, into a zone between 5°S and 15°N. Siberia continued to move north throughout the Palaeozoic, approaching and passing the North Pole in the Triassic.

It is interesting to note that the path of the Siberian plate from the Late Cambrian to the Recent lies along a great circle (Fig. 3). Such a simple movement suggests that the Siberian platform was part of a larger plate. It is not clear whether this plate comprised part of eastern Gondwana or southeast Asia, though the proximity of early Palaeozoic poles from Siberia and Gondwana has been noted (Khramov *et al.*, 1974; Bucha *et al.*, 1976; Khramov and Rodionov, 1977; Morel and Irving, 1978).

During the early and middle Palaeozoic the Russian plate occupied low latitudes. Through the Devonian and into the early Carboniferous it moved northward, rotating slightly clockwise. In the Tournasian and middle Carboniferous retrograde motion to the south took place.

The Late Proterozoic was a time of continental separation as Siberia, and probably Kazakhstan, rifted away from the Russian platform. In the early Palaeozoic the Russian and Siberian plates were much farther apart than they are at present. Both plates were brought together by oblique, northward, convergence throughout the Palaeozoic. This convergence was accompanied by transverse compression along the Uralian geosyncline. At about the same time, the Altai-Sayan fold belt was rotated clockwise and in the early Carboniferous was added to the Siberian plate. The Permian was the time of maximum convergence between the Russian and Siberian plates.

During the Permian Kazakhstan collided obliquely along the Urals. This collision was accompanied by the counterclockwise rotation of part of the Ural mountain belt. In the Triassic the northern edge of the Siberian platform moved slightly away from the Russian plate. The separation of the Taimyr peninsula and the formation of the Yensei-Khatanga rift system apparently took place at this time. Finally, the Turanian and Scythian plates, though subjected to internal deformation during the Palaeozoic and Mesozoic, were an integral part of the Russian platform by the beginning of the Mesozoic.

As it was mentioned above, the proposed model for the relative movement of the Siberian and Russian plates is but one possible solution. Alternate solutions can not be ruled out, though in light of the data presented here, they seem rather improbable. The problem of the late Precambrian and Palaeozoic motions of the Siberian and Russian platforms requires and warrants further study.

The authors are very grateful to R. A. Komissarova, E. L. Gurevich and S. A. Pisarevsky (VNIGRI) for palaeomagnetic determinations concerning the Proterozoic of Siberia, to B. Z. Asanidze, D. M. Pechersky (Institute of the Physics of the Earth) for useful comments on the Palaeozoic palaeomagnetism of the North Caucasus, to I. P. Slautsitais and E. P. Osipova (VNIGRI) for help during the preparation of the paper. The authors are especially grateful to R. E. Sorkina (institute of the Geology of the Arctic) for the English translation of the manuscript, and to Dr. C. Scotese who reviewed and edited this paper.

REFERENCES

ASANIDZE, B. Z. and D. M. PECHERSKY, Tectonicheskaya istoria otdelnykh blokov Kavkaza po paleomagnitnym dannym, Preprint N 4, 14 pp., Institut Fisiki Zemli, 1979.

BRIDEN, J. C., D. E. DREWRY, and A. G. SMITH, Phanerozoic equal area world maps, *J. Geol.*, **2**, 555–574, 1974.

BUCHA, V., Paleomagnetism and plate tectonics since the Lower Palaeozoic, *Studia Geophys. Geod.*, **19**, 42–60, 1975.

BUCHA, V., Z. MALKOVSKY, G. N. PETROVA, V. P. RODIONOV, K. ROTHER, and A. N. KHRAMOV, Problema kontinentalnogo dreifa na territorii Evrazii po dannym paleomagnitnykh issledovany nizhnego paleozoya, in *Dreif Kontinentov*, pp. 87–202, Nauka, Moskow, 1976.

CREER, K. M., A review of palaeomagnetism, *Earth Sci. Rev.*, **6**, 369–466, 1970.

FORSH, N. N. and A. N. KHRAMOV, Paleomagnetism i paleoclimaty na Russkoi platforme v karbone i permi, *Doklady Akad. Nauk*, **137**, 151–153, 1961.

GORODNITSKY, A. M., L. P. ZONENSHAIN, and E. G. MIRLIN, *Rekonstruktsii polozhenia materikov v fanerozoe*, 122 pp., Nauka, Moskow, 1978.

GUSEV, B. V., K voprosu o genezise mezokainozoiskikh vpadin zentralnoi chasti Sovetskoi Arktiki, in *Geofizicheskie metody razvedki v Arktike*, No. 7, pp. 68–81, Institut Geologii Arktiki, Leningrad, 1972.

GUSEV, B. V., Rekonstruktsia dokembriiskogo anomalnogo magnitnogo polya Russkoi i Sibirskoi platform, in *Geofizicheskie metody razvedki v Arktike*, No. 8, pp. 49–51, Institut Geologii Arktiki, Leningrad, 1973.

IRVING, E., E. TANCZYK, and J. HASTIE, Catalogue of paleomagnetic directions and poles, third issue. Palaeozoic results, Geomagn. ser N 5. Earth Physics Branch, Energy, Mines and Resources, Ottawa, 99 pp., 1976.

IVANOV, N. A., Ispolzovanie paleomagnithykh metodov dlya opredelenia nekotorykh elementov tektoniki gornykh porod Urala, in *Glubinnoe stroenie Urala*, pp. 52–54, Nauka, Moskow, 1968.

KHRAMOV, A. N. (Editor), *Paleomagnitnye napravlenia i paleomagnitnye polyusa*, Data for the USSR, Issue 1, 124 pp., VNIGRI, Leningrad, 1971.

KHRAMOV, A. N. (Editor), *Paleomagnitnye napravlenia i paleomagnitnye polyusa*, Data for the USSR, Issue 2, 88 pp., Materials of the WDC-B, Moskow, 1973.

KHRAMOV, A. N. (Editor), *Paleomagnitnye napravlenia i paleomagnitnye polyusa*, Data for the USSR, Issue 3, 43 pp., Materials of the WDC-B, Moskow, 1975.

KHRAMOV, A. N. (Editor), *Paleomagnitnye napravlenia i paleomagnitnye polyusa*, Data for the USSR, Issue 4, 51 pp., Materials of the WDC-B, Moskow, 1979.

KHRAMOV, A. N. and V. P. RODIONOV, Problema Lavrasii v rannem i srednem paleozoe v svete paleomagnitnykh dannykh. in *Paleomagnetism i voprosy tektoniki plit*, Trudy VNIGRI, issue 394, pp. 108–140, Leningrad, 1977.

KHRAMOV, A. N. and L. Ye. SHOLPO, *Paleomagnetism, Prinzipy metody i geologicheskie prilozhenia paleomagnitologii*, 251 pp., Nedra, Leningrad, 1967.

KHRAMOV, A. N., V. P. RODIONOV, and R. A. KOMISSAROVA, Novye dannye o paleozoiskoi istorii zemnogo magnitnogo polya na territorii SSSR, in *Nastoyashchee i proshloe magnitnogo polya Zemli*, pp. 206–213, Nauka, Moskow, 1965.

KHRAMOV, A. N. (Editor) *et al.*, *Paleomagnetism Paleozoja*, 236 pp., Nedra, Leningrad, 1974.

KIRSCHVINK, J. L., The Precambrian-Cambrian boundary problem: paleomagnetic directions from Amadeus basin, Central Australia, *Earth Planet. Sci. Lett.*, **40**, 91–100, 1978.

MCELHINNY, M. W., Palaeomagnetic directions and pole positions—XIII. Pole numbers 13/1 to 13/94, *Geophys. J. R. Astr. Soc.*, **30**, 281–283, 1972.

MCELHINNY, M. W., *Palaeomagnetism and Plate Tectonics*, 358 pp., Cambridge Univ. Press, Cambridge, 1973.

MCELHINNY, M. W. and J. A. COWLEY, Palaeomagnetic directions and pole positions—XIV. Pole numbers 14/1 to 14/574, *Geophys. J. R. Astr. Soc.*, **49**, 313–356, 1977.

MOREL, P. and E. IRVING, Tentative paleocontinental maps for the Early Phanerozoic and Proterozoic, *J. Geol.*, **86**, 535–561, 1978.

TAMRAZYAN, G. P., Siberian continental drift, *Tectonophysics*, **11**, 433–460, 1971.

The Concept of a Mobile Pangea and the Continuity of Continental Drift

P. Morel and E. Irving

Earth Physics Branch, Department of Energy, Mines and Resources, Ottawa, Canada

(Received February 2, 1980; Revised April 27, 1980)

Fifty-six years ago Koppen and Wegener (1924) published their map sequence depicting the displacement of the continents since the mid-Carboniferous about 320 million years ago. They envisaged a single supercontinent, Pangea, extending from the south pole to mid-northern latitudes. The last one-third of their map sequence, covering the last 100 million years, is familiar to everyone—the rupture of Pangea and the dispersal of the fragments forming the present continents. For the first two-thirds of their story Koppen and Wegener imagined that Pangea remained essentially unchanged for about 200 million years from the mid-Carboniferous to the Cretaceous. To these ideas of Koppen and Wegener must now be added those of Argand (1924) and later workers who have invoked extensive continental drift in the Early Palaeozoic prior to Koppen and Wegener's interval of no-drift in the later Palaeozoic and Mesozoic. Most workers who have written on this question seems to adhere, implicitly or explicitly, to this go-stop-go view of continental drift during the Phanerozoic (e.g. Smith *et al.*, 1973; Scotese *et al.*, 1979).

But is this picture of discontinuous continental drift really correct, or did the continents move relative to one another in a more-or-less continuous manner? In particular, is the concept of a fixed Pangea for Late Palaeozoic and Early Mesozoic time correct?

The question is important for four reasons. Firstly it is of basic importance to our understanding of Late Palaeozoic tectonics. Secondly, Carboniferous, Permian, and Triassic rocks contain large accumulations of coal, gas, and oil and it is important to understand the global tectonic framework of these deposits. Thirdly, it is important for the basic theory of palaeomagnetism, because as we show below, if the concept of a fixed Pangea for Late Palaeozoic and Triassic times is correct then it implies that palaeomagnetic pole positions obtained from all pre-Jurassic rocks are subject to generally undetectable and systematic errors of about 15°. Finally, the question is important because the competing concepts of discontinuous and continuous drift provide very different conditions for theories of mantle convection that seek to provide a mechanism for continental drift.

The mean palaeopoles obtained palaeomagnetically from Early Jurassic rocks of North America, Eurasia, and Gondwana when rotated with their respective land-masses

Earth Physics Branch Contribution No. 883.

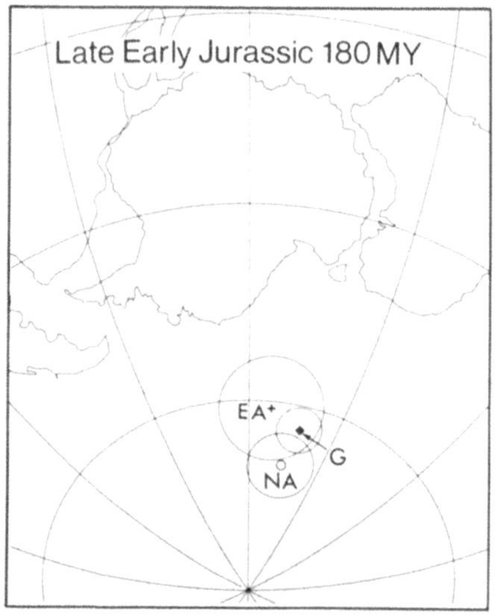

Fig. 1. Mean palaeopoles for North America (NA), Europe-northern Asia (EA), and Gondwana (G) plotted on the reconstruction of Pangea *A* using SMITH and HALLAM (1970) for Gondwana and LE PICHON *et al.* (1977) for the peri-Atlantic continents. Standard error circles are drawn.

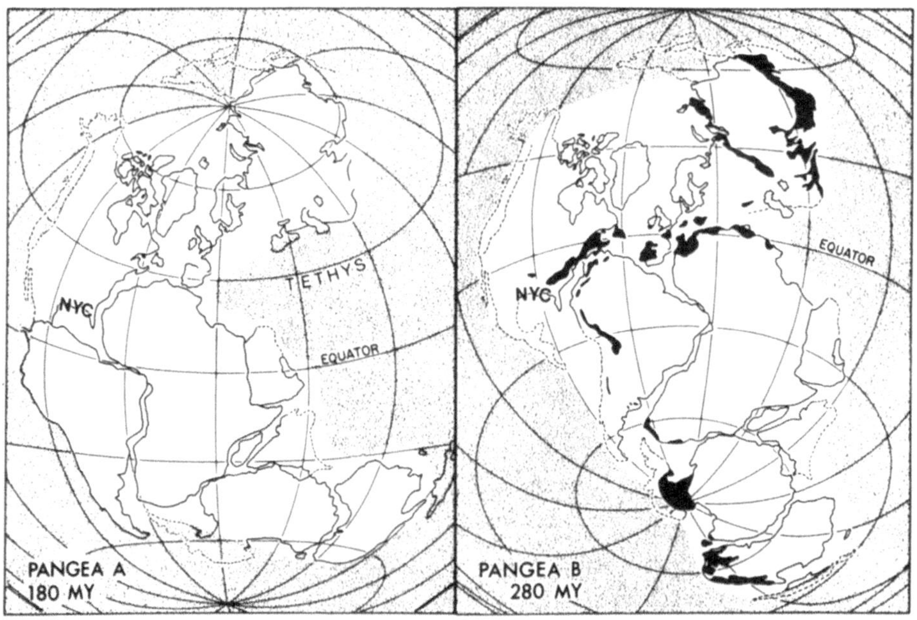

Fig. 2. Pangea *A* and *B* reconstructions based on palaeomagnetic evidence (MOREL and IRVING, 1981). Regions of Late Palaeozoic deformation (Hercynian, Appalachian orogenies etc.) are shown in black on Pangea *B*. NYC is the Nicaragua-Yucatan-Campeche Bark block of WHITE (1980).

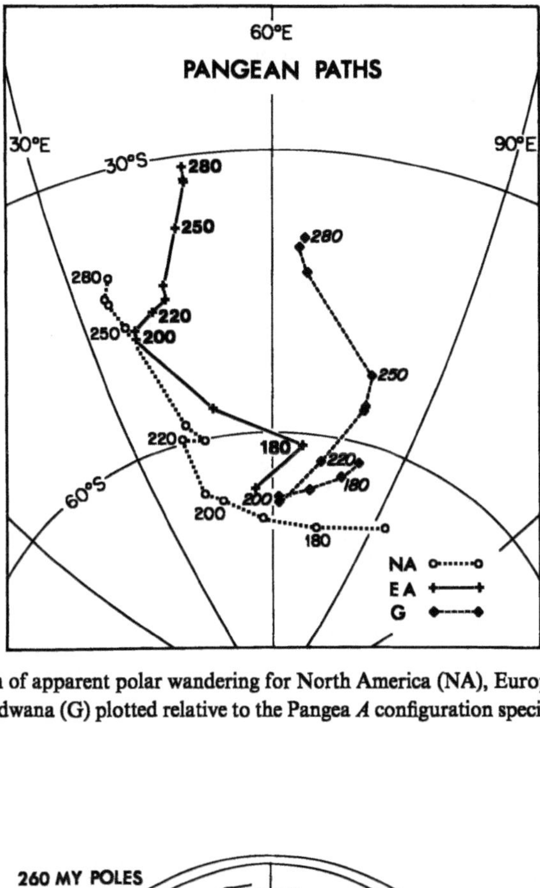

Fig. 3. Mean path of apparent polar wandering for North America (NA), Europe-northern Asia (EA), and Gondwana (G) plotted relative to the Pangea *A* configuration specified in the legend to Fig. 1.

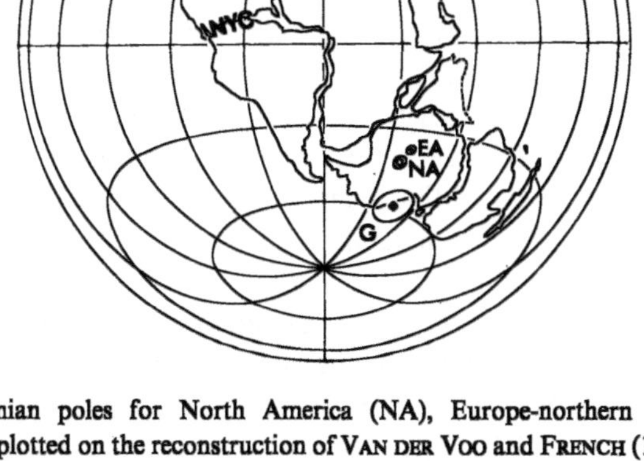

Fig. 4. Early Permian poles for North America (NA), Europe-northern Asia (EA), and Gondwana (G) plotted on the reconstruction of VAN DER VOO and FRENCH (1974). NYC is the Nicaragua-Yucatan-Campeche Bark block of WHITE (1980).

into Wegener's Pangea (we refer to configurations of this general type, e.g. that of BULLARD *et al.* (1965), as Pangea *A*) are in excellent agreement and confirm the general validity of the reconstuction for that interval of time (Fig. 1). Our preferred version of Pangea *A* is given on the left-hand side of Fig. 2. However if the analysis is extended backwards using data from older rocks the palaeopoles no longer agree (Fig. 3). This was first discovered by JAEGER and IRVING (1957, see IRVING, 1979). The divergences first become apparent in the late Triassic (200 m.y.), and are prominent in the Permian and Late Carboniferous (250 to 280 m.y.). Differences are present among all paths.

Is it possible to reconcile the Triassic and Late Palaeozoic palaeopoles by making some small readjustments of Wegener's Pangea yet retaining its essential outline for this interval of time? According to our analysis (MOREL and IRVING, 1981) the answer to this question is, no. The major discrepancy of Fig. 3, that between Gondwana and the northern continents in the Permian, requires that Gondwana be rotated clockwise. The rotation permissible within the general framework of Wegener's Pangea is shown Fig. 4. In this reconstruction, which is by VAN DER VOO and FRENCH (1974, Fig. 5), the northwest coast of South America is flush with the coast of the USA. Little or no further rotation is possible, if Africa is to be retained opposite to south-eastern North America. However the mean palaeopoles remain in disagreement; the palaeopoles for Eurasia (EA) and North America (NA) show a small divergence and the Gondwana palaeopole (G) is displaced to the south. The reconstruction of PILGER (1978) which is similar to that of Van der Voo and French also does not reconcile the palaeomagnetic data (MOREL and IRVING, 1980). Van der Voo and French based their Permian reconstructions of Gondwana relative to Laurasia on two palaeopole determinations from South America only (VAN DER VOO and FRENCH, 1974, Table 1, data of EMBLETON, 1970). VALENCIO (1972) and VALENCIO and MITCHEL (1972) suggest a Late Carboniferous age for the rocks from which these two palaeopoles were derived and VALENCIO *et al.* (1977, p. 71) have reaffirmed this opinion. There are now many more palaeopoles from Permian rocks from Africa and South America (15 palaeopoles) and from Gondwana as a whole (24 palaeopoles). All these palaeopoles are based on demagnetization studies. We find that Gondwana palaeopoles are systematically displaced relative to palaeopoles from North American and Europe (Fig. 3, MOREL and IRVING, 1981).

The scheme of VAN DER VOO and FRENCH (1974) for the Permian has been considered favourably because it provides a framework for the genesis of the Ouachita-Marathon orogenic belt (ROSS, 1979, Fig. 4) which is considered to have resulted from a Mid-Carboniferous continent-continent collision (WICKHAM *et al.*, 1976). However their reconstruction encounters difficulties in the east and in the west. In the east ARTHAUD and MATTE (1977, p. 1,319) have noted that such configurations for pre-Triassic times require "a large oceanic zone, the Tethys, extending along the southern edge of the European plate and more or less parallel to the southern Variscan front. This geometrical reconstruction is inconsistent with the characteristics of the Variscan Belt, which is typically the result of a collision between two continental plates." In the west, around the site of the modern Gulf of Mexico, WHITE (1980) had proposed that Nicaragua, Yucatan, and the Campeche Bank (NYC, Figs. 2 and 4) were situated immediately south of the Ouachita-Marathon orogenic belt in the Permian and Triassic and that they rifted from

North America in the Jurassic to form the Gulf of Mexico. White argues that his reconstruction explains the Triassic and Jurassic rifting events that occurred around the Gulf of Mexico and the distribution of the Louan salt. On his scheme it would have been the NYC mini-block that collided with North America to form the Ouachita-Marathon orogenic belt in the Carboniferous not NE South America. The reconstruction of Van der Voo and French provides no space for the NYC mini-block (Fig. 4). A reconstruction which we propose as a possible solution to this three-fold problem (the absence of an accurate fit to the palaeopoles, the absence of a southern margin to the Variscan orogenic belt of central Europe, and the absence of space for NYC mini-block) is shown on the right-hand side of Fig. 2. Africa is beneath Europe, northwestern South America is opposite the southern Appalachians, and the continents around what is nowadays the northern Atlantic are a little further apart than in Pangea *A*, producing a geologically acceptable rift between Scandinavia and Greenland. This configuration is called Pangea *B* and it is referred to the Late Carboniferous and Early Permian.

The reconstruction of Fig. 2 provides a symmetrical framework for the Hercynian (Variscan) foldbelt of Europe, placing it opposite the northwest Africa foldbelt, and also for the Appalachians placing them opposite Late Palaeozoic terranes of northwest South America. Space for NYC is provided. This concept implies that Pangea *A* (Wegener's Pangea) was a short-lived configuration lasting no more than a few tens of millions of years and was preceded in time by an interval of major right-lateral motion between Gondwana and the northern continents. Our estimate of the relative motion is about 3,500 km as a minimum. The transformation from Pangea *B* to *A* which, according to the palaeomagnetic data occurred between the Early Permian and Early Jurassic, presents many unresolved difficulties. Although there is good evidence of a post-Variscan dextral megashear between Gondwana and Laurasia the motions appear to be predominantly Late Carboniferous or Early Permian (ARTHAUD and MATTE, 1977) earlier that is than the motions implied palaeomagnetically (IRVING, 1977). Moreover there are serious problems with Florida. If southern Florida is to be regarded as part of Africa in the Palaeozoic which would seem to be commonly accepted (VAN DER VOO *et al.*, 1976) then the megashear between west Gondwana and North America must have been north of peninsular Florida. There is no evidence for such a megashear although it could be hidden beneath cover rocks of the Atlantic coastal plain.

The alternative viewpoint to that just given is that prior to the Jurassic the geomagnetic field had large non-dipole components with time constants of several tens of millions of years, and/or that the palaeomagnetically determined field directions are systematically in error by 10–20°. To accept this view is to say that all palaeopoles obtained from pre-Jurassic rocks may be subject to undetectable errors of the order of 15° so that palaeopole studies for these earlier times are crude and qualitative at best. In the Jurassic and later times these aberrations disappear, and the palaeomagnetically determined palaeopoles are in excellent accord with independently derived reconstructions (Fig. 1).

Obviously, the presently available palaeomagnetic evidence, taken in isolation, cannot provide a conclusive answer to this intriguing and longstanding problem. The problem will yield when very many more palaeomagnetic data become available, especially from northern South America and northern Africa, and when the tectonic and stratigraphic history of the Appalachian-Hercynian superfoldbelt is reviewed in its entirety. Nevertheless the evidence as it stands at present indicates that continental drift could have been a continuous process not a discontinuous one, as envisaged by KOPPEN and WEGENER (1924), SCOTESE *et al.* (1979), and others. The tectonically distinguishing feature of Late Palaeozoic and Triassic motions is that they appear to have produced predominantly lateral slip between the continental blocks rather than ocean-floor spreading and subduction that appears to be such a characteristic feature of Early Palaeozoic and of later Mesozoic and Cenozoic motions.

REFERENCES

ARGAND, E., Tectonique de l'Asie, Session XIII, Liege 1922, *C. R. Congr. Géol. Inter.*, **1**, 171–372, 1924.

ARTHAUD, F. and P. MATTE, Late Paleozoic strike-slip faulting in southern Europe and northern Africa: Result of a right-lateral shear zone between the Appalachians and the Urals, *Geol. Soc. Am. Bull.*, **88**, 1305–1320, 1977.

BULLARD, E. C., J. E. EVERETT, and A. G. SMITH, The fit of the continents around the Atlantic, in A symposium on continental drift, *Phil. Trans. R. Soc., Lond., Ser. A*, **258**, 41–51, 1965.

EMBLETON, B. J. J., Paleomagnetic results for the Permian of South America and a comparison with the African and Australian data, *Geophys. J. R. Astr. Soc.*, **21**, 105–118, 1970.

IRVING, E., Drift of the major continental blocks since the Devonian, *Nature*, **270**, 304–309, 1977.

IRVING, E., Pole positions and continental drift since the Devonian, in *The Earth: its Origin, Structure and Evolution*, edited by M. W. McElhinny, pp. 567–593, Academic Press, 1979.

JAEGER, J. C. and E. IRVING, Palaeomagnetism and the reconstruction of Gondwanaland, *C. R. 3rd Congr. Pacific Indian Ocean Sci. A*, Tananarive, 233–242, 1957.

KOPPEN, W. and A. WEGENER, *Die Klimate der geolischen Vorzeit*, Borntrager, Berlin, 1924.

LE PICHON, X., J. C. SIBUET, and J. FRANCHETEAU, The fit of the continents around the north Atlantic ocean, *Tectonophysics*, **38**, 169–209, 1977.

MOREL, P. and E. IRVING, Late Paleozoic reconstruction of the continents based on paleomagnetism, in *The Origin of the Gulf of Mexico and the Early Opening of the Central North Atlantic Ocean*, edited by R. H. Pilger, Jr., pp. 75–78, School of Geoscience, Louisiana State Univ., Baton Rouge, 1980.

MOREL, P. and E. IRVING, Paleomagnetism and the evolution of Pangea, *J. Geophys. Res.*, 1981 (in press).

PILGER, R. H., Jr., A closed Gulf of Mexico, pre-Atlantic ocean plate reconstruction and the early rift history of the Gulf and North Atlantic, *Transactions—Gulf Coast Association of Geological Societies*, XXVIII, 385–393, 1978.

ROSS, C. A., Late Paleozoic collision of North and South America, *Geology*, **7**, 41–44, 1979.

SCOTESE, C. R., R. K. BAMBACH, C. BARTON, R. VAN DER VOO, and A. M. ZIEGLER, Paleozoic base maps, *J. Geol.*, **87**. 217–277, 1979.

SMITH, A. G. and A. HALLAM, The fit of the southern continents, *Nature*, **225**, 139–144, 1970.

SMITH, A. G., J. C. BRIDEN, and G. E. DREWRY, Phanerozoic world maps, in *Organisms and Continents through Time*, edited by N. F. Hughes, *Palaeontology, Special Papers*, **12**, 1–42, 1973.

VALENCIO, D. A. Intercontinental correlation of late Palaeozoic South American rocks on the basis of their magnetic remanences, *Ann. Acad. Brasil. Cienc., Suppl.*, **44**, 357–364, 1972.

VALENCIO, D. A. and J. G. MITCHELL, Palaeomagnetism and K-Ar ages of Permo-Triassic igneous rocks from Argentina and the inter-continental correlation of Upper Palaeozoic—Lower Mesozoic formations, 24th Int. Geol. Congress, Section 3, pp. 189–195, 1972.

VALENCIO, D. A., J. F. A. VILAS, and J. E. MENDIA, Paleomagnetism of a sequence of redbeds of the Middle and Upper Sections of Paganzo Group (Argentina) and the correlation of Upper Palaeozoic—Lower Mesozoic rocks, *Geophys. J. R. Astr. Soc.*, **51**, 59–74, 1977.

VAN DER VOO, R. and R. B. FRENCH, Apparent polar wandering for the Atlantic-bordering continents: Late Carboniferous to Eocene, *Earth Sci. Rev.*, **10**, 99–119, 1974.

VAN DER VOO, R., F. J. MAUK, and R. B. FRENCH, Permian-Triassic continental configurations and the origin of the Gulf of Mexico, *Geology*, **4**, 177–180, 1976.

WHITE, G. W., Permian-Triassic continental reconstruction of the Gulf of Mexico-Caribbean area, *Nature*, **283**, 823–826, 1980.

WICKHAM, J., D. ROEDER, and G. BRIGGS, Plate tectonics models for the Ouachita foldbelt, *Geology*, **4**, 173–176, 1976.

Model of the World Plate Tectonics since the Early Palaeozoic

Václav BUCHA

Geophysical Institute of Czechoslovak Academy of Sciences, Prague, Czechoslovakia

(Received March 29, 1980)

The pattern of palaeomagnetic curves with particular regard to the lower Palaeozoic epochs which was revised by a suitable choice of polarity of data for the individual continents is given. The investigated distribution of the continental blocks for the individual geological epochs agrees with all the palaeontological and geological facts. The comparison of the model of continental drift with the occurrence of orogenic events and some palaeontological conclusions was carried out.

1. Intorduction

Thanks to the efforts of hundreds of specialists abundant material and data on palaeomagnetic poles for the individual geological epochs were accumulated through intensive palaeomagnetic research over several decades (RUNCORN, 1960; IRVING *et al.*, 1976; KHRAMOV, 1974; MCELHINNY, 1973; CREER, 1970, and others). Nevertheless, the interpretations of the results obtained, in particular as regards the Lower and Middle Palaeozoic, were not fully successful and generated a great deal of doubt about the validity of the obtained palaeomagnetic poles, so that specialists from various fields, particularly some geologists, are still not fully convinced of the validity of some palaeomagnetic results.

According to TEICHERT and MEYERHOFF (1972) "the evidence disproves the widely held and widely believed concepts of a Pangaea, Laurazia, and Gondwanaland with a vast continental interior covered with continental deposits."

MEYERHOFF (1970): "Today evaporites are deposited in two belts 10°–50° north and 10°–38° south of the equator. The results show that ancient evaporites were deposited as far north as 83°N and as far south as 39°S. However, palaeomagnetic studies have not helped to explain the presence of evaporites at high latitudes; they only make explanation more difficult. For example, if Irving's latitudes are used, the latitudinal spread of evaporites is from 80°N to 71°S.

The spread of "reliable" poles on each continent is wider than the Atlantic Ocean where drift is supposed to have taken place. How, then, can the method be used to determine relative motions between continents?"

According to MEYERHOFF (1970), Martin also doubted the reliability of palaeomagnetic data for Africa and wrote that the placing of a Devonian pole on Africa is contradicted by the extensive development of Devonian coral reefs in West Africa and the presence of Devonian evaporites in Ghana and North Africa.

According to MEYERHOFF (1970), "several palaeomagneticians arrived at conclusions which are unsupported by modern meteorology, climatology, physical oceanography, and plant geology.

This concerns also the distribution of late Palaeozoic glacial centres and middle Palaeozoic to Tertiary coal deposits. Those who advocate drift and/or polar wandering must provide a satisfactory explanation for these tillite and coal distributions to prove that new concepts of global tectonics are valid".

I think we should not ignore these and other comments but we should look for their explanation because only a positive discussion can improve our understanding with geologists.

2. Palaeomagnetic Polar Wandering Curves

The known curves for individual continents are given in Fig. 1a, b. We can see a big deviation of the Ordovician pole for North America and England as well as a big distance between the Ordovician and Silurian pole for Australia. Further (Fig. 2), the comparison of Ordovician results for North America (RUNCORN, 1960), Czechoslovakia (BUCHA, 1961, 1969), and Siberia (KHRAMOV, 1974) are given. The declinations for all three regions (130°–145°) do not differ considerably, the dip is between +25° for North America to −25° for Czechoslovakia and −10° for Siberia. The Ordovician pole positions for North America, Europe, and Siberia were determined (Fig. 3a). After rotating, and shifting North America and Europe in such a way that all three poles coincided (see Figs. 3a, b, c) we can consider such a configuration as given in Fig. 3c to have existed in the past. When we rotate it in order to make the Ordovician paleopole coincide with the present geographic pole we obtain also the original orientation in the Ordovician (Fig. 3d). Our data from Czechoslovakia (BUCHA, 1961, 1969) showed a comparatively unusual pattern

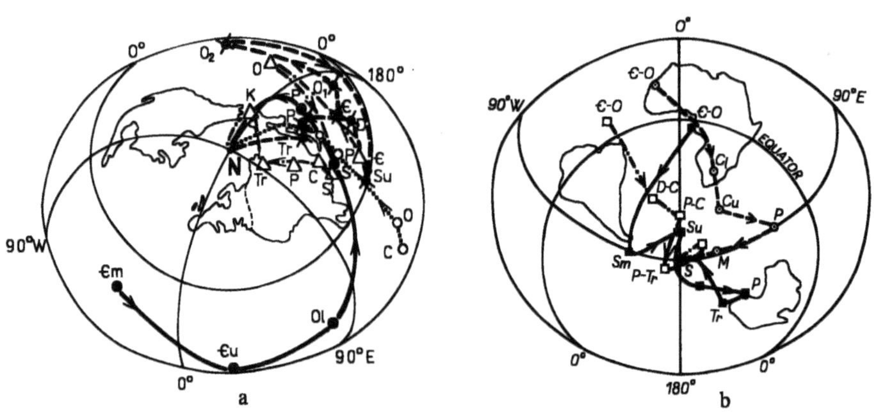

a b

Fig. 1. Palaeomagnetic curves expressing the apparent wandering of the north pole for the interval between the Cambrian and the present: a) ●—— for Czechoslovakia (BUCHA, 1961), ○···· for Siberia (KHRAMOV, 1974), △·-·-·-· for North America, ✖------ England, b) ⊙------ Africa, ■—— Austalia, □-·-·-·-· South America (MCELHINNY, 1973).

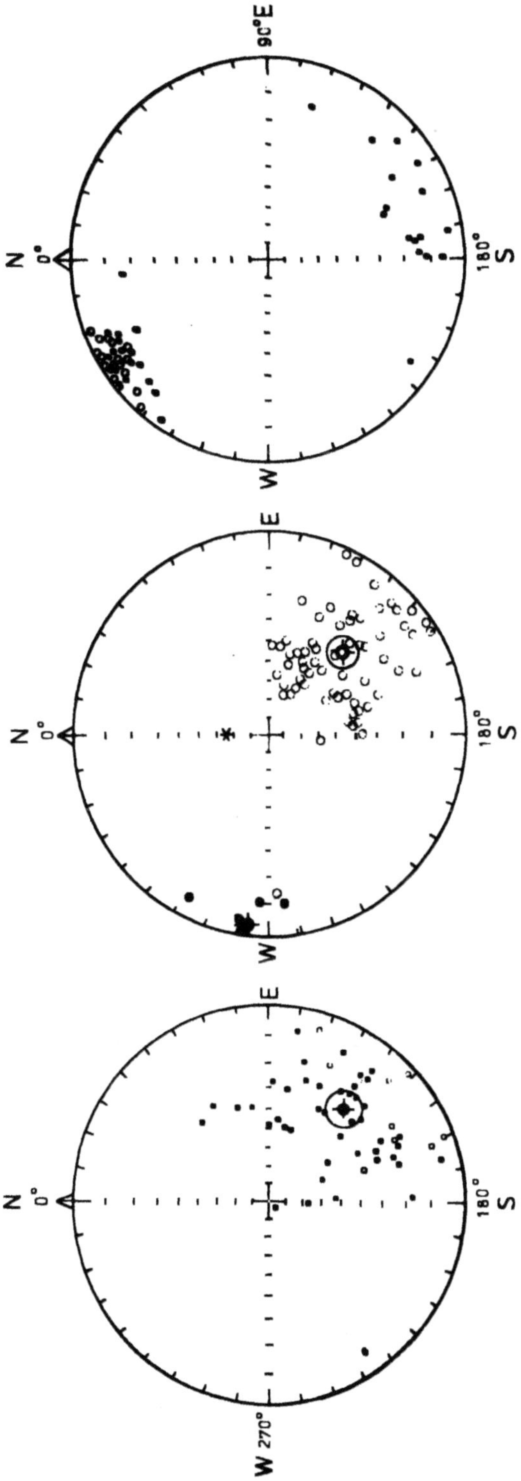

Fig. 2. Plot of directions of magnetization of the Ordovician rocks from North America (Runcorn, 1960)—left-hand side, Czechoslovakia (Bucha, 1961, 1969)—in the middle, Siberia (Khramov, 1974)—right-hand side.

Fig. 3. Schematic configuration of the continental plates (North America, Europe, Asia) and the position of Ordovician poles. a) recent configuration, b) when identifying the North American with the European pole, c) when identifying the North American and European pole with the Asian (Siberian) pole, d) after identifying the joint Ordovician paleopole with the present geographical pole and after a simultaneous rotation of all the three continental plates. The data given in Fig. 2 were used to express the pole positions for North America, Europe (Czechoslovakia), Asia (Siberia) in the Ordovician.

in that they displayed the opposite direction to some of the curves then known for Europe and North America. For this reason we checked some of the results which were decisive for the pattern of the curve, particularly as regards the Lower Ordovician and the Cambrian (BUCHA, 1969). It was found that with an exception of a few smaller deviations, the original pattern of the curve remained unchanged, i.e. for the Cambrian it passes through Central America towards the east, then through South Africa into the Indian Ocean (Lower Ordovician), and from here towards Japan (Silurian) and then to the north (Fig. 4).

Our results from the territory of Czechoslovakia (BUCHA, 1961, 1965, 1969) indicated a possibility of showing that there existed a single ancient continent, the Pangaea, in the Lower Palaeozoic, including the Asian and North American continents in the overall configuration of all continents. However, one must bear in mind that the palaeomagnetic (averaged) poles representing the position of the geographic poles, are two in number and that the occurrence of reverse polarity, so frequent in the history of the geomagnetic field, prompts one to be careful in deciding which of the poles to adopt in constructing the palaeomagnetic curves of the individual blocks.

As regards the curve derived for Australia (Fig. 4), we considered all the early Palaeozoic data (MCELHINNY, 1973) separately, i.e. early, middle, late Cambrian, early, late Ordovician, and not just the mean value; in this way, the big spread of Cambrian and Ordovician poles as objected by Meyerhoff can be reduced considerably. It is known that the Cambrian and Ordovician represent a time interval of more than 160 million years and therefore, it does not seem very probable that Australia and North America would have remained "put" in a single place over this long span of time. Then, the palaeomagnetic polar-wandering curve of the North Pole for Australia proceeds similarly to the cs curve, only slightly more to the north across present Africa (Cambrian–Ordovician) towards Siberia (Silurian) and then to the north. We consider the poles for the Cambrian and Ordovician to be north, so that the "polar-wandering" curve has a quite different pattern and is longer in the early Palaeozoic. However, in this connection the distance between the Ordovician and Silurian poles is shorter than in the connection given earlier.

The curve for Siberia, constructed from Soviet data (KHRAMOV, 1974) is then presented (Fig. 4).

The polar-wandering curve for North America (Fig. 4) displays an even longer trajectory than the European curve, determined by Czechoslovak palaeomagnetic data. Among other reasons, this is also due to the continent departing by 40° from Europe and Africa over the past 150 million years as a result of the relatively large speed of wandering of both Americas in the past.

Also the curve for England showing a similar pattern to the curve for North America is given (Fig. 4).

The set of curves for the 6 continental blocks (Fig. 4) indicates (BUCHA, 1975a, b) that they differ completely from the curves, hitherto published (Fig. 1). The Cambrian and Ordovician poles, located in the region of the present Atlantic and Africa were also considered as north for the southern continents (Australia, Africa), i.e. again with a polarity opposite to that employed by other authors. Moreover, it is evident that the distances between Cambrian and Silurian poles for Australia, Europe, Africa were shorter than those considered earlier.

The palaeomagnetic curves as given here have approximately the shape of spirals, converging on today's North Pole. The pattern of the African curve where Silurian values were not available—these should be considered tentative and should be shifted in such a way that Carboniferous pole coincided with the recent pole. It is interesting that none of the curves passes through the region of the present Mediterranean and North Africa, so that one may assume that this region is located approximately in the common centre of

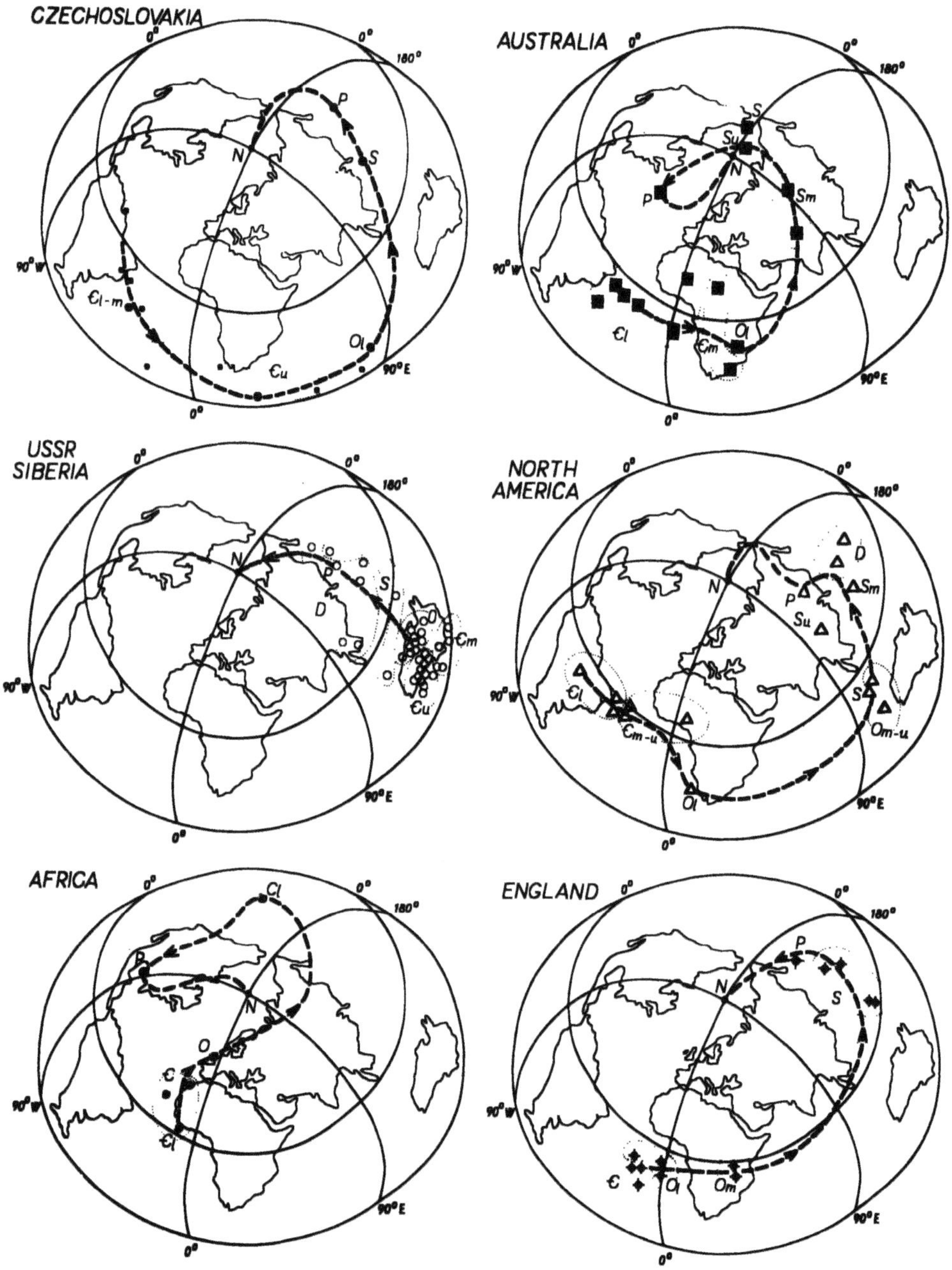

Fig. 4. Set of new palaeomagnetic curves, expressing the apparent wandering of the north palaeomagnetic (geographic) pole for the interval between the Cambrian and the recent for Czechoslovakia, Australia, USSR (Siberia), North America, Africa, England: C—Cambrian, O—Ordovician, S—Silurian, D—Devonian, C—Carboniferous, P—Permian. Cambrian and Ordovician poles were taken with opposite polarity for Australia, North America, Africa and England to those given by McElhinny (1973).

the whole set of curves mentioned (30°N, 30°E). This fact seems to be important for the further interpretation of the results obtained.

3. Palaeomagnetic Model of the Continental Drift since the Cambrian

In order to obtain an idea of the overall arrangement of the continental blocks and of their development in the course of the drift over the Earth's surface, the present shapes of the continents were not divided into more parts than the number of present continents. It is known, for example, that the European continent was divided into two or more parts (north and south), and that the Asian continent could similarly have been divided into several parts (north, east, south).

It was considered that the common centre of the curves was located in the present northeastern Africa and individual continents were gradually displaced by rotating them around this centre for all 12 periods since the Cambrian to the present (Bucha, 1975a, b).

This distribution of the continents, determined for 12 intervals of time since the Cambrian to the present, expresses their configuration on the present Earth's surface as determined from the palaeomagnetic data in the present geographic network (Fig. 5).

It should be pointed out that this is a tentative rough configuration of the continents constructed by means of a graphical method of transformation, the more so because for some continents data from all geological epochs considered were not available; more accurate results are expected when computational methods are applied.

Figure 5 shows a similar angular velocity but different rates of motion of the continents; the most marked difference is between North America and Europe. Whereas North America moved 25,000 km over 600 million years, i.e. approximately 4–6 cm per year, Europe rotated with a velocity of 1–2 cm per year (approx. 9,000 km over 600 million years), Asia (Siberia) with a velocity of 2 cm per year. These data agree with the rate of the seafloor spreading determined by other methods.

It seems that the original ancient continent was located eccentrically in the region between the North Pole and the equator, and it, therefore, displayed a tendency to move towards the equator as a result of Eötvös's force. Because of this the Coriolis force began to act on the continent, westwards in the Northern Hemisphere; the ancient continent began to rotate clockwise. Its parts more remote from the centre of rotation (30°N, 30°E) had a tendency to move faster than the parts closer to the centre, and the continent gradually began to disintegrate, as can be seen in Fig. 5.

The continental drift is conditioned by the existence of the asthenosphere extending from a depth of 80 to 300 km in the Earth's mantle and characterized as a low velocity channel of seismic waves. The temperature of the mass here reaches almost the melting point and its viscosity is from two to three orders lower than that of the ambient layers. This magnitude (approx. 10^{21} poise) makes the mass behave as a rigid liquid towards the longterm acting forces. The lithospheric continental blocks, which display horizontal movements of several cm per year as a result of the above mentioned forces float on the asthenosphere.

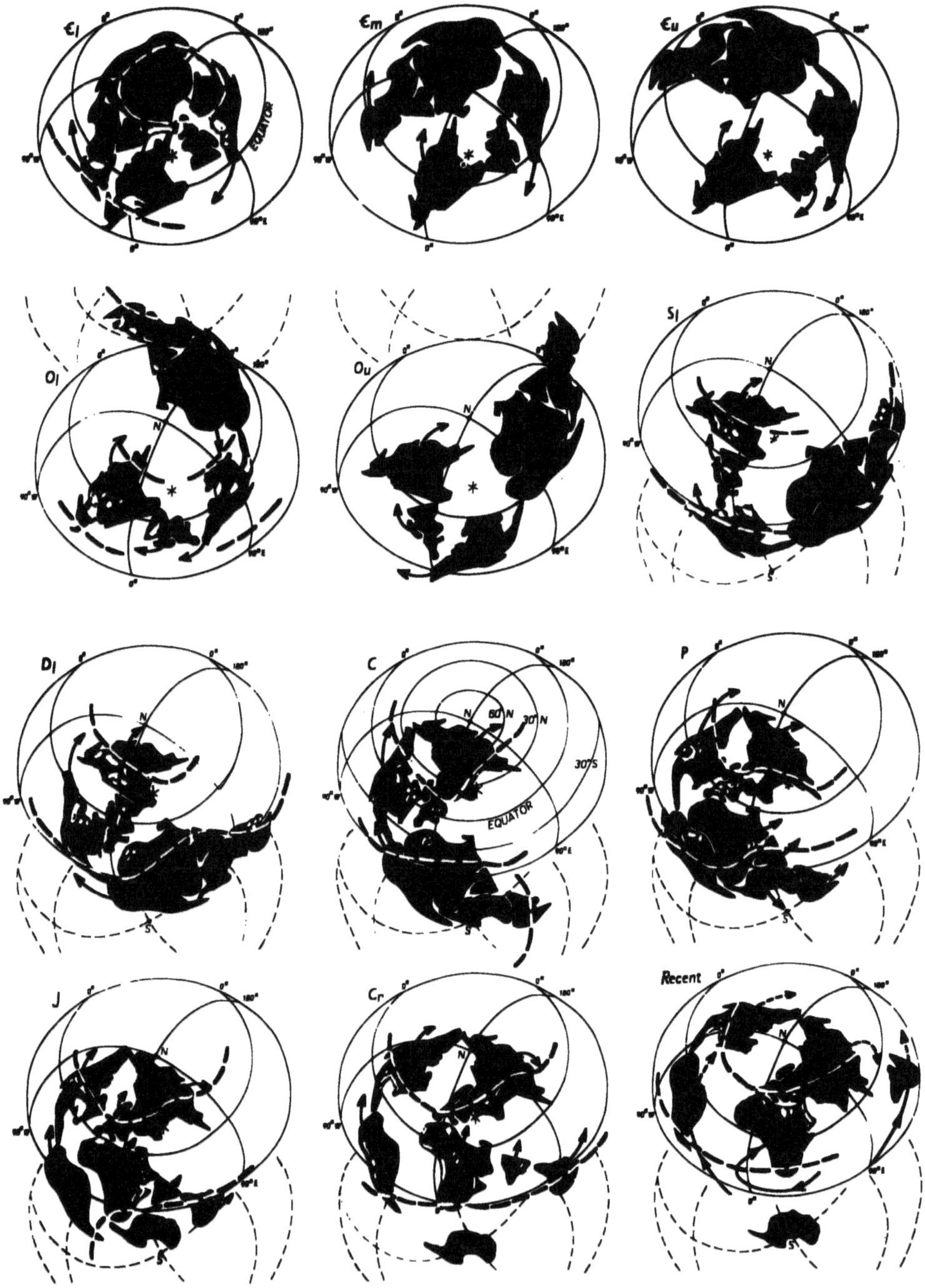

Fig. 5. Schematic configuration of the continental plates on the Earth's surface in the present geographic network: C_l—lower Cambrian, C_m—middle Cambrian, C_u—upper Cambrian, O_l—lower Ordovician, O_u—upper Ordovician, S_l—lower Silurian, D_l—lower Devonian, C—Carboniferous, P—Permian, J—Jurassic, C_r—Cretaceous. The arrows indicate the direction of motion of the individual continents and * denotes the approximate common centre of rotation around which most of the continents rotated. The distribution of ancient evaporites for individual geological periods is marked as white areas and dotted curves (MEYERHOFF, 1970).

4. Continental Drift and Orogenic Processes

The proposed system of development of tectonic blocks allows one to express certain ideas of the possible causes of the orogenes on the Earth's surface. We know that the orogenic processes in the individual geological periods displayed different intensities over the past 600 million years as can be seen from Fig. 6.

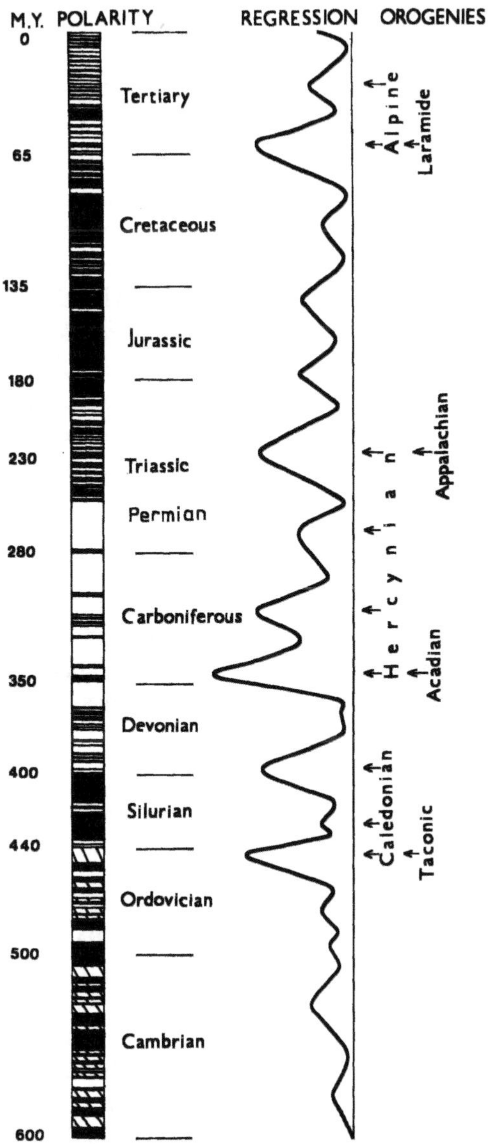

Fig. 6. Reversals of the polarity of the Earth's magnetic field and orogenic events measured as percent regression from a smooth curve passed through the peaks of marine transgression (DAMON, 1971).

The comparison of the model of continental drift (Fig. 5) with the occurrence of orogenesis (Fig. 6) implies the following conclusions (BUCHA, 1975b).

In the Cambrian and Ordovician, the Pangaea gradually opened up and the "southern" continents were shifted into the region of the present Pacific Ocean. No significant collisions between the continents occurred at the time so that the orogenic activity during these periods was also relatively low.

During the Upper Ordovician and the Silurian North America collided with Europe and this was reflected in the strong Caledonian folding.

At the end of the Devonian and in the Carboniferous the southern continents approached the Eurasian block, Europe collided with Africa from the south and tectonic processes carried on as a result of the faster shift of North America along Europe from its western side. Figure 5 implies that the southern continents and North America moved together between the Cambrian and Permian. The result of the extensive collisions between the Lower Carboniferous and the Upper Permian was the intense Hercynian folding, the creation of the Ural and other orogeneses.

In the Upper Devonian North America collided with the north-eastern part of the Asian continent and this was responsible for the considerable deceleration of the motion of North America and of the southern continents, and also affected the intense Hercynian orogenesis, terminated in the Lower Triassic. These extensive tectonic processes also apparently had a decisive effect on the disintegration of the southern plate into several smaller tectonic units and, together with other forces (Eötvös and Coriolis) caused Africa, India and Australia to shift from the South Pole towards the north-east at a relatively high speed after the separation.

No very significant collisions between the continental blocks occurred during the Mesozoic and, therefore, also the orogenic activity was relatively low.

After the approach and collision of moving Africa and India with the slowly rotating Eurasian continent at the end of the Cretaceous and the beginning of the Tertiary, the very marked Alpine, Caucasian and Himalayan foldings took place along the whole Tethys zone.

5. Comparison of the Proposed Palaeomagnetic Model with Palaeontological Data since the Cambrian

The proposal for the model of the dynamic system of the world-wide plate tectonics and of its developments, presented here and in BUCHA (1975a, b) agrees with all palaeontological facts about the occurrence of warm climate (coral reefs, evaporites, coal deposits) and about the glaciations.

According to PALMER (1972), the present distribution of Cambrian trilobite fauna is inconsistent with a static concept of world geography during Phanerozoic time (Fig. 7). During the Cambrian, Mexico and its faunas of European affinities are at the high latitude end, typical American faunas across the United States are in the intermediate latitudes, and Alaska is in lower latitudes (Fig. 8). This complies with our model of tectonic blocks in the Cambrian (see the position of North America in Fig. 5).

Now to the objection of MEYERHOFF and TEICHERT (1971) that the new global

Fig. 7. Map of the continental plates showing occurrences of fossiliferous Cambrian rocks and distribution of restricted shelf margin-open ocean regions (PALMER, 1972).

Fig. 8. North America showing relationship of paleolatitudes to distribution of restricted shelf faunas and shelf margin-open ocean faunas in the Cambrian (PALMER, 1972).

tectonics fails completely to explain the distribution of late Palaeozoic glacial centers and middle Palaeozoic coal deposits.

Our model shows that the southern continents were not always at the same place during the Carboniferous and Permian, but moved from the South Pole to the northwest (see Figs. 5 and 9). At that time the Antarctica moved from the South Pole, together with the whole Gondwanaland by approximately 30° northwards. Further one has to take into consideration that the configuration of the southern continents was different than that given by MEYERHOFF (1970), particularly as regards the position of the Antarctica. That's why the South Pole was not in the centre of the Gondwanaland either but on its border (see Fig. 9); this means that the major glacial centres in the Carboniferous were not too far inland and enough moisture could have reached them (Fig. 9).

Coal deposits could not originate on the territory of Antarctica and Australia during

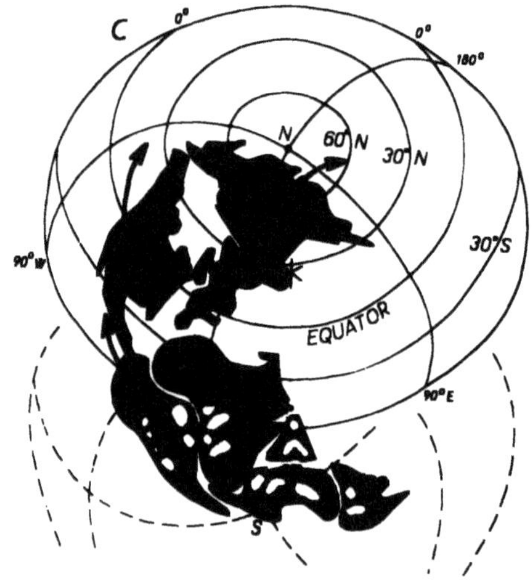

Fig. 9. Distribution of glacial centers in the Carboniferous (MEYERHOFF, 1970).

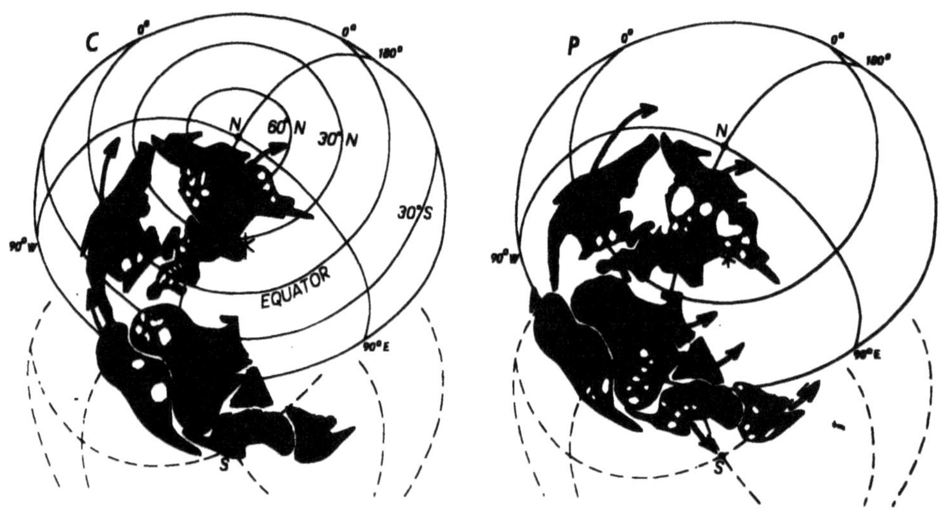

Fig. 10. Distribution of coal basins. C—in the Carboniferous, P—in the Permian (MEYERHOFF, 1970).

the Carboninferous because both these continents were near the South Pole at that time (Fig. 10). On the other hand, in the Permian, after the shift of both continents towards the equator, favourable conditions for coal deposits were created—in the Permian both continents had average geographical latitude of 45°–60°S. Glacial centres originated mainly in the Carboniferous. The Carboniferous continental glaciation and the Permian mountain glaciation could be mingled and intermixed.

The distribution of receptaculitids indicating the position of the tropical belt in the (Lower) Palaeozoic (Fig. 11) conflicts according to NITECKI (1972) with the prevailing palaeogeographic interpretation of the position of the equator and the poles relative to

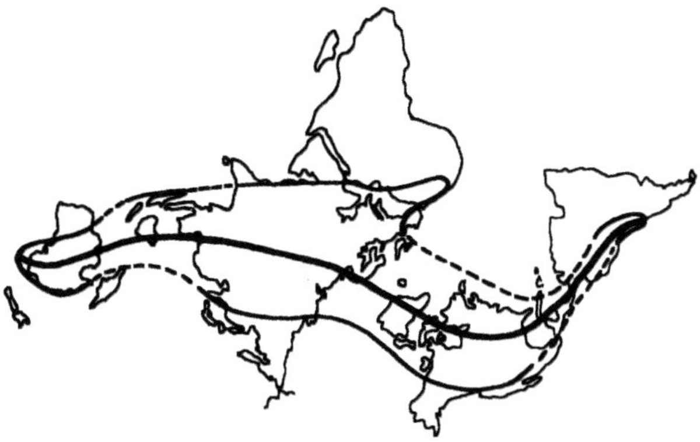

Fig. 11. The geographic distribution of receptaculitids. The assumed position of the palaeo-equator is indicated by the heavy line (NITECKI, 1972).

the continents during the Palaeozoic.

Our model can also explain all these conclusions (compare Fig. 5 with Figs. 7 and 11) because the continents of Australia, Northern Asia, the Antarctica, the western part of North and South America and Europe were either on the equator or close to it during the Lower and Middle Palaeozoic.

According to TEICHERT and MEYERHOFF (1972) "the evidence disproves the widely held and widely believed concepts of a Pangaea, Laurazia and Gondwanaland. The few faunal-realm studies which have been made show clearly that Asia has been connected to North America via the Bering Strait, and probably via the Novaya Zemlya-Lomonosov Ridge and Spitzbergen-Greenland, since the late Devonian. The conclusion is demonstrated by the fact that the faunal realms of Europe grade eastward and northward into those of northwestern Asia and North America".

In considering our configuration of continents in the Devonian (Fig. 5) we can see that it complies with Meyerhoff's and Teichert's conclusions. Our model shows that Pangaea existed in the Cambrian (and also in the Precambrian); it disproves, however, the concept of Laurazia and Gondwanaland, as presented in some older palaeomagnetic models.

The direction of motion of the individual continents towards the pole near Hawai in our model complies with Jeffreys' conclusions.

Our model also explains the ancient distribution of evaporites by the gradually changing configuration of the continents as given in Fig. 5. This correlation of our model with the evaporite data shows that their distribution corresponds to the belts along the equator for all geological periods as given by MEYERHOFF (1970).

Thus, we can see that this interpretation of the palaeomagnetic data, (Fig. 5), can answer and explain practically all palaeontological and palaeclimatological data.

There are not enough palaeomagnetic data for the Ordovician and Silurian which seem to be decisive for the definite course of the polar wandering curves, mainly for Europe, South and North America, Africa and the Antarctica. Palaeomagneticians should concentrate their efforts on this important question, the solution of which could help considerably in explaining the configuration of the continents in these geological periods in more detail.

REFERENCES

BUCHA, V., Palaeomagnetic Pole Positions in the Pre-Cambrian and Palaeozoic periods investigated from Czechoslovak rocks, *Studia Geophys. Geod.*, **5**, 269–273, 1961.

BUCHA, V., Interpretation of Results of Palaeomagnetic Research, Travaux Inst. Géophys. Acad. Tchécosl. Sci., No. 206, Geof. sborník 1964, Academia, Praha, 1965.

BUCHA, V., Palaeomagnetism of the Upper Cambrian and the Lower Ordovician in Czechoslovakia, Travaux Inst. Géophys. Acad. Tchécosl. Sci., No. 296, Geof. sborník 1968, Academia, Praha, 1969.

BUCHA, V., Palaeomagnetism and Plate Tectonics since the Lower Palaeozoic, *Studia Geophys. Geod.*, **19**, 42–59, 1975a.

BUCHA, V., Palaeomagnetism and the Continental Drift, *Studia Geophys. Geod.*, **19**, 300–305, 1975b.

CREER, K. M., A Review of Palaeomagnetism, *Earth Sci. Rev.*, **6**, 369–466, 1970.

DAMON, P. E., *The Relationship between Late Cenozoic Volcanism and Tectonism and Orogenic—Epeirogenic Periodicity*, Conference on "The Late Cenozoic Glacial Ages," edited by K. K. Turekian, pp. 15–31, John Wiley & Sons, New York, 1971.

IRVING, E., E. TANCZYK, and J. HASTIE, *Catalogue of Paleomagnetic Directions and Poles*, Third Issue, Paleozoic Results 1949–1975, Energy, Mines and Resources Canada, Geomagnetic Series, No. 5, Ottawa, 1976.

KHRAMOV, A. N., *Paleomagnetizm Paleozoa*, Izd. Nedra, Leningrad, 1974.

MCELHINNY, M. W., *Palaeomagnetism and Plate Tectonics*, Cambridge Univ. Press, Cambridge, 1973.

MEYERHOFF, A. A., Continental Drift: Implications of Paleomagnetic Studies, Meteorology, Physical Oceanography and Climatology, *J. Geology*, **78**, 1–51, 1970.

MEYERHOFF, A. A. and C. TEICHERT, Continental Drift, III: Late Paleozoic Glacial Centres, and Devonian—Eocene Coal Distribution, *J. Geology*, **79**, 285–321, 1971.

NITECKI, M. H., The Paleogeographic Significance of Receptaculitids, Intern. Geol. Congr. XXIV sess., Sec. 7, Paleontology, Montreal, pp. 303–307, 1972.

PALMER, A. R., Problems of Cambrian Biogeography, Intern. Geol. Congr., XXIV sess., Sec. 7, Paleontology, Montreal, pp. 310–315, 1972.

RUNCORN, S. K., Polar Wandering and Continental Drift: Evidence from Paleomagnetic Observations in the Unites States, *Bull. Geol. Soc. Am.*, **71**, 915–930, 1960.

TEICHERT, C. and A. A. MEYERHOFF, Continental Drift and the Marine Environment, Intern. Geol. Congr., XXIV sess., Sec. 7, Paleontology, Montreal, pp. 339–349, 1972.

A Computer Animation of Continental Drift

Christopher R. Scotese,* Sigmund Snelson,** William C. Ross,**
and Linda P. Dodge**

*Department of Geophysical Sciences, University of Chicago,
Chicago, Illinois, U.S.A.
**Shell Development Co., Bellaire Research Center,
Houston, Texas, U.S.A.

(Received May 12, 1980)

A computer animation of continental drift has been produced which illustrates the movements of the continents during the last 545 million years. Beginning in the Late Cambrian, the animation shows the sequence of continental collisions which by the end of the Palaeozoic formed the supercontinent, Pangaea. The remainder of the film describes the breakup of Pangäea and the subsequent formation of the world's ocean basins. The animation was made at Shell Development Company, Houston, Texas. It is entitled, *Dynamic Continents*, and is black and white, silent, and runs approximately 5 min.

1. Introduction

During the past 10 years, the tectonic history of nearly every mountain belt and ocean basin has been re-interpreted in light of plate tectonic theory. Detailed maps of linear magnetic anomalies and data from the Deep Sea Drilling Project permit accurate reconstructions of the spreading histories of all the major ocean basins. These recent plate tectonic models, together with the framework provided by palaeomagnetism, have been used by various workers to reassemble the past configurations of the continents.

The first set of widely accepted Phanerozoic reconstructions were made by Smith *et al.* (1973). These maps, which showed the positions of the continents at 8 intervals since the Late Cambrian, have served as base maps for numerous geologic, biogeographic, and palaeoclimatic investigations. Recently, the Mesozoic and Cenozoic reconstructions have been revised and expanded (Smith and Briden, 1977). Other reconstructions of the Palaeozoic, Mesozoic, and Cenozoic have been proposed by Zonenshayn and Gorodnitsky (1977a, b), Kanasewich *et al.* (1978), Morel and Irving (1978), Irving (1977), and Scotese *et al.* (1979). Though differing in detail, these models outline the same basic pattern of Palaeozoic collision and Mesozoic-Cenozoic rifting.

Though the maps are useful for plotting and interpreting a variety of geologic, biogeographic, and palaeoclimatic data, they cannot capture the dynamic aspects of the plate tectonic process. In plate tectonics, subtle temporal relationships are often as important as the more obvious spatial relationships. The best way of adding the dimension of time to a set of reconstructions is by film animation.

In 1974 a film based on Smith *et al.*, (1973) set of reconstructions was produced on the PLATO computer system at the University of Illinois in Chicago (SCOTESE and BAKER, 1975). As a result of the success of the first film, work was begun on a new animation at Shell Development Company's Bellaire Research Center. The film described in this article is one of several animations which have been produced during the past two years.

2. Materials and Methods

The computer hardware used to generate the film consists of a Harris minicomputer, Tektronix 4014 terminal, Versatec electrostatic plotter, and Adage vector display terminal. A Bolex, 16 mm motion picture camera was interfaced with the system and the animation was filmed directly from the Adage screen.

The Adage is an interactive graphics terminal. Built-in microprocessors permit such 'hard-wired' graphics functions as 3-dimensional rotations, scaling and windowing. Hard copy of any screen display can be made using the Versatec electrostatic plotter.

Because the Adage cathode ray tube constantly updates, or refreshes, the vector display, rather than permanently store the vector positions, animation sequences can be previewed or filmed in 'real time'. The slowest link in the animation process is the 16 mm camera which films at a rate of two frames per sec. In a typical session, 3 min of animation could be filmed in a little less than one hour.

Though the film appears to be continuous it is actually composed of 24 separate animated sequences. Versatec plots of 6 frames from the animation are shown in Figs. 1–6. The data used to construct this model of Phanerozoic plate motions are described in the following section.

3. The Plate Tectonic Model

Though the overall pattern of continental motion during the Phanerozoic is known, there are times and places where crucial information is not available, or where the interpretation of data is disputed. A film does not provide a suitable format for the discussion of alternate points-of-view, nor is there any way of placing 'error bars' on the interpretations shown. In the following section, the plate tectonic model and the palaeomagnetic data which were used to produce the animation will be discussed and areas of controversy outlined.

4. The Palaeozoic

The palaeomagnetic data and the Euler rotations used to reassemble the Palaeozoic continents are described in SCOTESE *et al.* (1979). The pattern of Palaeozoic continental movements, collisions, and palaeogeographic changes is outlined in two companion papers (ZIEGLER *et al.*, 1979; BAMBACH *et al.*, 1980).

The Palaeozoic reconstructions are based primarily on palaeomagnetic data. Though absolute longitude cannot be determined, the relative longitudinal arrangement can in most cases be deduced from biogeographic patterns (ZIEGLER *et al.*, 1977), and the sequence of continental collisions. As Fig. 1 illustrates, the problem of longitudinal uncertainty was minimized during the early Palaeozoic for this was a time when most of the continents occupied the same low latitudinal belt.

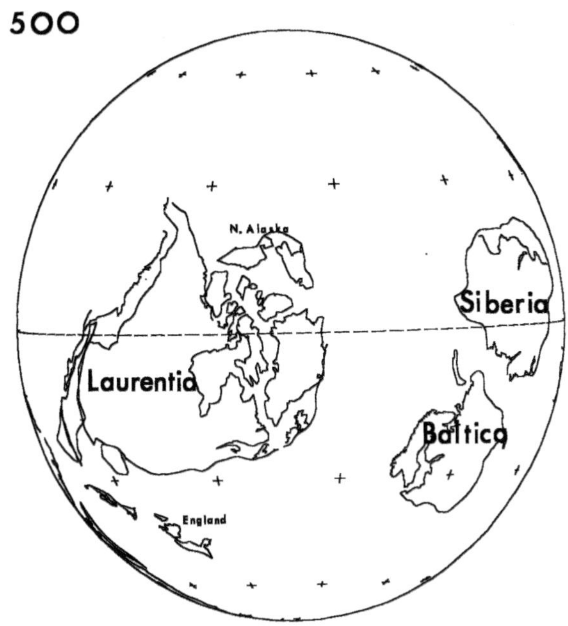

Fig. 1. Early Ordovician (500 million years ago).

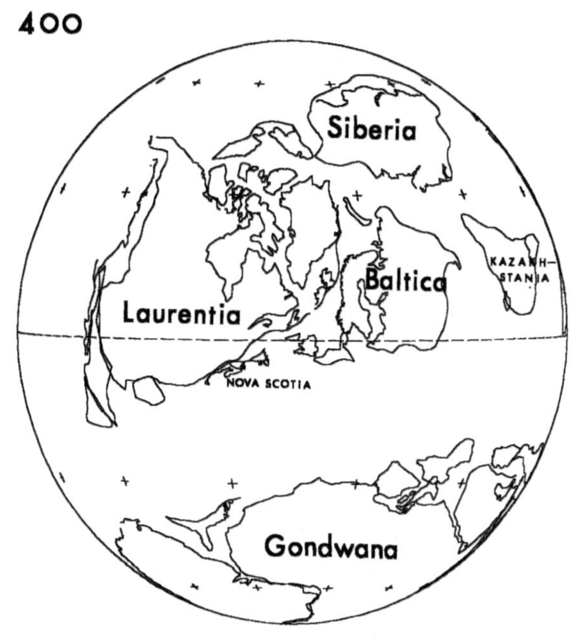

Fig. 2. Early Devonian (400 million years ago).

Fig. 3. Late Carboniferous (300 million years ago).

Reliable palaeomagnetic data for the early Palaeozoic are available from North America (Laurentia), Siberia, and Gondwana. Cambro-Ordovician data from Europe suggest that England and the Baltic shield were subtropical, while parts of southern and central Europe (Armorica) were associated with Gondwana near the South Pole (HAGSTRUM *et al.*, 1980). There is no palaeomagnetic data from southeast Asia; however, broad carbonate platforms and faunal affinities indicate an equatorial association with Australia.

During the Middle Palaeozoic the oceans separating these continents were subducted and the first of several continental collisions took place (Fig. 2). The Taconic, Caledonian, and Acadian orogenies have all been variously attributed to continent-continent collision (BIRD and DEWEY, 1970; MCKERROW and ZIEGLER, 1972; VAN DER VOO, 1979). Though the timing of the collisions and the areas involved are disputed, all interpretations recognize that by the Middle Devonian Laurentia and Baltica had collided to form the 'Old Red continent'.

Recent palaeomagnetic determinations from the Late Devonian of North America (KENT and OPDYKE, 1978; VAN DER VOO, 1979) indicate that the arrangement of the continents after the collisions was not the same as the configuration at the time of breakup in the Jurassic. Baltica originally collided in a more southerly position relative to Laurentia (compare Figs. 2 and 3). This orientation was maintained from Late Silurian through Early Carboniferous (KENT, 1980; KENT and OPDYKE, 1979).

Throughout the early and middle Palaeozoic, Gondwana continued to move northward. The width of the ocean separating Gondwana from the northern continents is not known because there are few reliable palaeomagnetic determinations from the southern continents of Siluro-Devonian age.

200

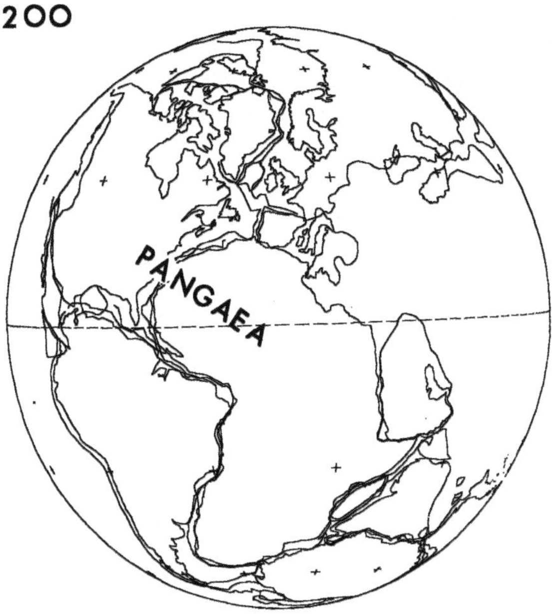

Fig. 4. Late Triassic (200 million years ago).

An alternate Devonian reconstruction has been proposed (Heckel and Witzke, 1979) which eliminates this southern ocean. In this interpretation, North America, Europe, and Siberia are displaced 30 degrees to the south so that they are adjacent to the northern coast of Africa. Though this 'Middle Palaeozoic Pangaea' explains the distribution of a variety of climatic indicators and accounts for the observed similarity between North American and South American faunas it can not be reconciled with numerous late Devonian palaeomagnetic determinations from North America and Siberia.

A single Silurian pole from China (Khramov, 1975) places southeast Asia in equatorial latitudes during the Middle Palaeozoic. A few determinations from the middle and late Devonian of Kazakhstan suggest that this area of southwestern U.S.S.R. occupied latitudes intermediate between Baltica and Siberia.

By the end of the Palaeozoic the formation of Pangaea was nearly completed. Though numerous Permo-Carboniferous palaeomagnetic determinations are available, there is considerable controversy regarding how this data should be used to reconstruct Pangaea.

Two versions of Pangaea have been proposed, Pangaea (A) and Pangaea (B). Pangaea (A) is the standard Wegnerian reconstruction (Fig. 4). Pangaea (B), in contrast, places North America against the northern coast of South America (Irving, 1977). Irving suggests that Late Palaeozoic palaeomagnetic data require that the collision of the northern and southern continents resulted in the configuration, Pangaea (B). This relationship was maintained from early Carboniferous through early Triassic (Irving, 1977). During the middle and late Triassic, Pangaea (B) was transformed to Pangaea (A) by means of strike slip motion between Gondwana and the northern continents. The concept of Pangaea (B) has yet to be critically tested using tectonic, geologic or palaeogeographic data. A preliminary calculation of the rates of relative motion during the transition from Pangaea (B) to Pangaea (A) indicates that during the Triassic Gondwana moved past North America at a relatively rapid rate of 15 cm/year. An analysis of the distribution of Permian fusulinids suggests that the faunal provinces are better explained by Pangaea (A) (ROSS, 1979).

An alternate interpretation of late Permian and early Mesozoic palaeomagnetic data does not require such drastic readjustments (VAN DER VOO and FRENCH, 1974; VAN DER VOO *et al.*, 1976). In the 'Van der Voo fit', North America is rotated tightly against northwest Africa, closing the Gulf of Mexico.

In the animation, Gondwana and the northern continents are shown shearing past one another during the early stages of collison. The Carboniferous is a time of intense tectonic activity both in Maritime Canada and Great Britain. Readjustments along the former Baltica-Laurentia collision zone took the form of left lateral transcurrent faulting which resulted in the opening of grabens in the Gulf of St. Lawrence and the North Sea. ARTHAUD and MATTE (1976) indicate that there is evidence of right lateral strike slip motion along the southern margin of the Baltic Shield which also may be associated with the collision of Gondwana.

5. The Mesozoic and Cenozoic

The Mesozoic and Cenozoic portions of the animation are based on recent plate tectonic models of the Atlantic (SCLATER *et al.*, 1977) and Indian oceans (NORTON and SCLATER, 1979). Palaeomagnetic data from North America were used to orient the global reassemblies with respect to the pole (VAN ALSTINE and DE BOER, 1978).

In general, from Cretaceous through to the Recent, the plate motions shown in the film follow the outline of continental movements described by SMITH and BRIDEN (1977) in their set of Mesozoic and Cenozoic palaeocontinental maps. The differences in the two models are reviewed elsewhere (SCOTESE, 1980). Prior to the Cretaceous, there are uncertainties regarding the early breakup history of the southern continents, and the tectonic histories of the Gulf of Mexico, Mediterranean, Arctic, and Antarctic regions.

The early spreading history of the Indian Ocean is not known, the there is still considerable debate as to how the eastern and western halves of Gondwana must be assembled. In Du Toit's original reconstruction of the southern continents, Madagascar was placed in the northern notch along the eastern coast of Africa, that is, adjacent to Kenya. Palaeozoic palaeomagnetic poles cluster most tightly when this fit is used (KLOOTWIJK, 1978). It has been proposed that the alternate fit of Madagascar adjacent to Mozambique results in a better alignment of the Tasman, Trans-Antarctic, and Cape fold belts (BARRON *et al.*, 1978).

A third reconstruction of the Indian Ocean (JOHNSON *et al.*, 1980), which though similar to Du Toit's fit, displaces India approximately 500 km to the south. This reconstruction matches Jurassic shorelines in Africa, Madagascar and India, and may better explain the pattern of magnetic anomalies in the Mozambique Basin. The reconstruction shown in the animation links the eastern and western halves of Gondwana together using a modified SMITH and HALLAM (1970) fit (Fig. 4).

The early tectonic history of the Gulf of Mexico is similarly constrained by the 'fit' used to reconstruct the surrounding continental blocks. The reconstruction used in the film requires that Mexico, Honduras, and the Yucatan peninsula be rotated to the

100

Fig. 5. Middle Cretaceous (100 million years ago).

50

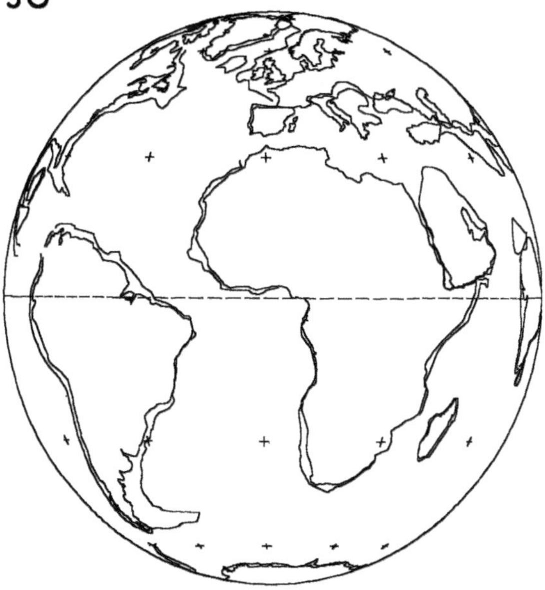

Fig. 6. Eocene (50 million years ago).

northwest and out of the way of South America. There is some evidence for major sinistral faults in Mexico and southwestern United States along which this motion may have taken place (SILVER and ANDERSON, 1974; MCKEE and JONES, 1979). Other reconstructions of the Gulf of Mexico displace Mexico and Central America in a similar manner (PILGER, 1978).

The tectonic histories of the Arctic and Antarctic regions must remain speculative at this time. Palaeomagnetic evidence may support a counterclockwise rotation of the northern peninsula of Alaska away from Arctic Canada (SWEENEY *et al.*, 1978). The timing of this rotation is not documented, but it may have coincided with some of the thrusting in the Brooks Range (TAILLEUR and SNELSON, 1969). Portions of eastern Siberia must also be included with this northern Alaskan block, however, it is not clear where the western suture lies.

Western Antarctica is probably composed of several microcontinents. In addition to the Western Antarctic peninsula, the Ellsworth mountains and Marie Byrdland must be accounted for in any reconstruction of the southern oceans. The reassembly shown in the film is similar to the reconstruction of DE WIT (1977). Marie Byrdland is collapsed against East Antarctica, and the Ellsworth mountains are rotated clockwise so that the folded Palaeozoic terrains line up with the main trend of the Trans-Antarctic mountains. The Western Antarctic peninsula is fitted against the Falkland plateau, providing a southern continuation of the Andean Cordillera. An alternate reconstruction based on palaeomagnetic data (ALLEY and WATTS, 1979) rotates the peninsula clockwise against eastern Antarctica, closing the Weddell Sea.

6. Conclusion

The computer animation described in this article allows the viewer to evaluate the dynamic and temporal aspects of continental motions. The film is naturally limited by the decreasing quality of geologic information the further one goes back in time. It is planned that as new data becomes available, and as new technology develops, we will be able to make improvements in the animation.

REFERENCES

ALLEY, R. and D. R. WATTS, Paleomagnetic investigation of the northern peninsula, *EOS, Trans. Am. Geophys. Union*, **60**, 240, 1979.

ARTHAUD, F. and P. MATTE, Late Paleozoic strike-slip faulting in southern Europe and northern Africa: Results of the right-lateral shear zone between the Appalachians and the Urals, *Geol. Soc. Am. Bull.*, **88**, 1305–1320, 1977.

BAMBACH, R. K., C. R. SCOTESE, and A. M. ZIEGLER, Before Pangea: the geography of the Paleozoic world, *Am. Scientist*, **68**, 26–38, 1980.

BARRON, E. J., C. G. A. HARRISON, and W. W. HAY, A revised reconstruction of the southern continents, *EOS, Trans. Am. Geophys. Union*, **59**, 436–449, 1978.

BIRD, J. M. and J. F. DEWEY, Lithosphere plate: Continental margin tectonics and the evolution of the Appalachian orogen, *Geol. Soc. Am. Bull.*, **81**, 1031–1059, 1970.

DE WIT, M. J., The evolution of the Scotia Arc as a key to the reconstruction of southwestern Gondwanaland, *Tectonophys.*, **37**, 53–81, 1977.

HAGSTRUM, J. T., R. VAN DER VOO, B. AUVRAY, and N. BONHOMMET, Eocambrian-Cambrian palaeomagnetism of the Armorican massif, France, *Geophys. J. R. Astr. Soc.*, **60**, 489–517, 1980.

HECKEL, P. H. and B. J. WITZKE, Devonian world paleogeography determined from the distribution of carbonates and related lithological paleoclimatic indicators, Special Papers in Palaeontol., No. 23, pp. 99–123, 1979.

IRVING, E., Drift of the major continental blocks since the Devonian, *Nature*, **270**, 304–309, 1977.

JOHNSON, B. D., C. MCA. POWELL, and J. J. VEEVERS, Early spreading history of the Indian Ocean between India and Australia. Earth Plan, *Sci. Lett.*, 1980 (in press).

KANASEWICH, E. R., J. HAVSKOV, and M. E. EVANS, Plate tectonics in the Phanerozoic, *Can. J. Earth Sci.*, **15**, 919–955, 1978.

KENT, D. V., The formation of Euramerica in the Paleozoic, *EOS, Trans. Am. Geophys. Union*, **61**, 220–221, 1980.

KENT, D. V. and N. D. OPDYKE, Paleomagnetism of the Devonian Catskill red beds: evidence for motion of coastal New England—Canadian maritime region relative to cratonic North America, *J. Geophys. Res.*, **83**, 4441–4450, 1978.

KENT, D. V. and N. D. OPDYKE, Further evidence for the early Carboniferous field of North America and its bearing on Carboniferous tectonics of the northern Appalachians, *Earth Planet. Sci. Lett.*, **44**, 365–372, 1979.

KLOOTWIJK, C. T., A review of paleomagnetic data from the Indo-Pakistani fragment of Gondwanaland, in *Geodynamics of Pakistan*, edited by A. Favah and K. A. DeJong, pp. 41–80, Pakistan Geol. Survey Pub., 1978.

KHRAMOV, A. N., Palaeomagnetic directions and paleomagnetic poles, Data for U.S.S.R., in *Materials for World Data Center B*, Vol. 3, Soviet Geophysical Committee of the Academy of Sciences of the U.S.S.R., 1975.

MCKEE, J. W. and N. W. JONES, A large Mesozoic fault in Coahuila, Mexico, Geol. Soc. Am., Abstracts with programs, Vol. 11, p. 476, 1979.

MCKERROW, W. S. and A. M. ZIEGLER, Palaeozoic Oceans, *Nature*, **240**, 92–94, 1972.

MOREL, P. and E. IRVING, Tentative paleocontinental maps for the early Phanerozoic and Proterozoic, *J. Geol.*, **86**, 535–561, 1978.

NORTON, I.O. and J. G. SCLATER, A model for the evolution of the Indian Ocean and the breakup of Gondwanaland, *J. Geophys. Res.*, **84**, 6803–6830, 1979.

PILGER, R. H., A closed Gulf of Mexico, pre-Atlantic Ocean plate reconstruction and the early rifting history of the Gulf of Mexico and the North Atlantic, *Trans. Gulf Coast Assoc.*, **28**, 385–393, 1978.

ROSS, C. A., Late Paleozoic collision of North and South America, *Geology*, **7**, 41–44, 1979.

SCLATER, J. G., S. HELLINGER, C. TAPSCOTT, The paleobathymetry of the Atlantic Ocean from the Jurassic to the present, *J. Geol.*, **85**, 509–552, 1977.

SCOTESE, C. R. and D. W. BAKER, Continental drift reconstructions and animation, *J. Geol. Educ.*, **23**, 167–171, 1975.

SCOTESE, C. R., R. K. BAMBACH, C. BARTON, R. VAN DER VOO, and A. M. ZIEGLER, Paleozoic base maps, *J. Geol.*, **87**, 217–277, 1979.

SCOTESE, C. R., A review of Mesozoic and Cenozoic Paleocontinental Maps, by A. G. Smith and J. C. Briden, New York, 1977 (Cambridge Univ. Press), *Am. J. Sci.*, **280**, 93–96, 1980.

SILVER, L. T. and T. H. ANDERSON, Possible left-lateral early to middle Mesozoic disruption of the southwestern North American craton, Geol. Soc. Am., Abstracts with programs, Vol. **6**, p. 955, 1974.

SMITH, A. G., J. C. BRIDEN, and G. E. DREWRY, Phanerozoic world maps, in *Organisms and Continents Through Time*, Palaeontol. Assoc. Spec. Pap., **12**, 1–42, 1973.

SMITH, A. G. and J. C. BRIDEN, *Mesozoic and Cenozoic Paleocontinental Maps*, 63 pp., Cambridge Univ. Press, New York, 1977.

SMITH, A. G. and A. HALLAM, The fit of the southern continents, *Nature*, **225**, 139–144, 1970.

SWEENEY, J. F., E. IRVING, and J. W. GEUER, Evolution of the Arctic Basin, in *Arctic Geophysical Review*, Vol. 45, edited by J. F. Sweeney, pp. 91–100, Pub. Earth Physics Branch, Ottawa, Canada, 1978.

TAILLEUR, I. L. and S. SNELSON, Large scale thrusting in northwestern Alaska possibly related to rifting of the

Arctic Ocean. (abs)., Geol. Soc. Am. Spec. Pap., **121**, p. 569, 1969.

VAN ALSTINE, D. R. and J. DE BOER, A new technique for constructing apparent polar wander paths and the revised Phanerozoic path for North America, *Geology*, **6**, 137–139, 1978.

VAN DER VOO, R. and R. B. FRENCH, Apparent polar wandering for the Atlantic bordering continents: Late Carboniferous to Eocene, *Earth Sci. Rev.*, **10**, 99–119, 1974.

VAN DER VOO, R., F. J. MAUK, and R. B. FRENCH, Permian—Triassic continental configuration and the origin of the Gulf of Mexico, *Geology*, **4**, 177–180, 1976.

VAN DER VOO, R., A. N. FRENCH, and R. B. FRENCH, A paleomagnetic pole position from the folded late Devonian catskill redbeds, and its megatectonic implications, *Geology*, **7**, 345–348, 1979.

VAN DER VOO, R., Paleozoic assembly of Pangaea: a new model for the Taconic Caledonian, and Hercynian orogenies, *EOS, Trans. Am. Geophys. Union*, **6**, 241, 1979.

ZIEGLER, A. M., C. R. SCOTESE, W. S. MCKERROW, M. E. JOHNSON, and R. K. BAMBACH, Paleozoic Paleogeography, *Ann. Rev. Earth Planet. Sci.*, **7**, 473–502, 1979.

ZIEGLER, A. M., C. R., SCOTESE, W. S. MCKERROW, M. E. JOHNSON, and R. K. BAMBACH, Paleozoic biogeography of the continents bordering the Iapetus (pre-Caledonian) and Rheic (pre-Hercynian) Oceans, in *Paleontology and Plate Tectonics, Special Pub. in Bio. and Geo.*, No. 2, pp. 1–34, Milwaukee Public Museum, 1977.

ZONENSHAYN, L. P. and A. M. GORODNITSKIY, Paleozoic and Mesozoic reconstructions of the continents and oceans, Article 1.—Early and Middle Paleozoic reconstructions, *Geotectonics*, **11**, 83–94, 1977a.

ZONENSHAYN, L. P. and A. M. GORODNITSKIY, Paleo-Mesozoic and Mesozoic reconstructions of the continents and oceans, Article 2.—Late Paleozoic and Mesozoic reconstructions, *Geotectonics*, **11**, 159–172, 1977b.

Palaeomagnetism of Lower Ordovician and Upper Precambrian Rocks from Argentina

Daniel A. VALENCIO

Departamento de Ciencias Geológicas, Universidad de Buenos Aires,
Ciudad Universitaria, Pabellón 2, Buenos Aires, Argentina

(Received December 13, 1979)

Palaeomagnetic data from samples collected in a rock sequence of tuffs and shales (Suri Formation), exposed in northwestern Argentina, indicate that the geomagnetic field had reversed polarity in Llanvirnian time. The samples from the top of the sequence also record, at least, an excursion of the geomagnetic field.

The palaeomagnetic study carried out on samples collected in a rock sequence of shallow marine sediments of Late Precambrian age (La Tinta Formation), exposed in central-eastern Argentina, indicates that the geomagnetic field had alternating periods of normal and reversed polarity in that geological time.

The positions of the palaeomagnetic poles for the Suri and La Tinta Formations are consistent with the interpretation that Gondwana was a unit as far back as Late Precambrian and with the occurrences of Upper Ordovician and Upper Precambrian glacial deposits across Africa and Australia, respectively. They do not, however, exclude models involving the opening and closing of minor seas. Particularly, the position of La Tinta pole is consistent with the interpretation which suggests that the sediments of this formation and those of the Nama Group (Africa) were deposited in a common shallow marine basin.

1. Introduction

The palaeomagnetic study of samples collected in a rock sequence yields data which may be used to define the behaviour of the geomagnetic field and the position of the collecting site relative to the pole.

Palaeomagnetic data for a sequence of Lower Ordovician rocks from Argentina (Suri Formation; VALENCIO *et al.*, 1980a) are used to define the characteristics of the geomagnetic field and the position of South America relative to the other gondwanic continents in Early Ordovician time.

DU TOIT (1927) suggested that the non fossiliferous marine sediments exposed in the southern parts of South American and Africa belong to the same Late Precambrian system. DALLA SALDA (1979) pointed out that the sediments included in La Tinta Formation (Late Precambrian), exposed in central Argentina, and the sediments of the lower part of the Nama Group (Late Precambrian to Cambrian), exposed in southern Africa, were accummulated in a common shallow-marine basin. Recently, palaeomagnetic data for a sequence of Upper Precambrian sediments from La Tinta Formation were

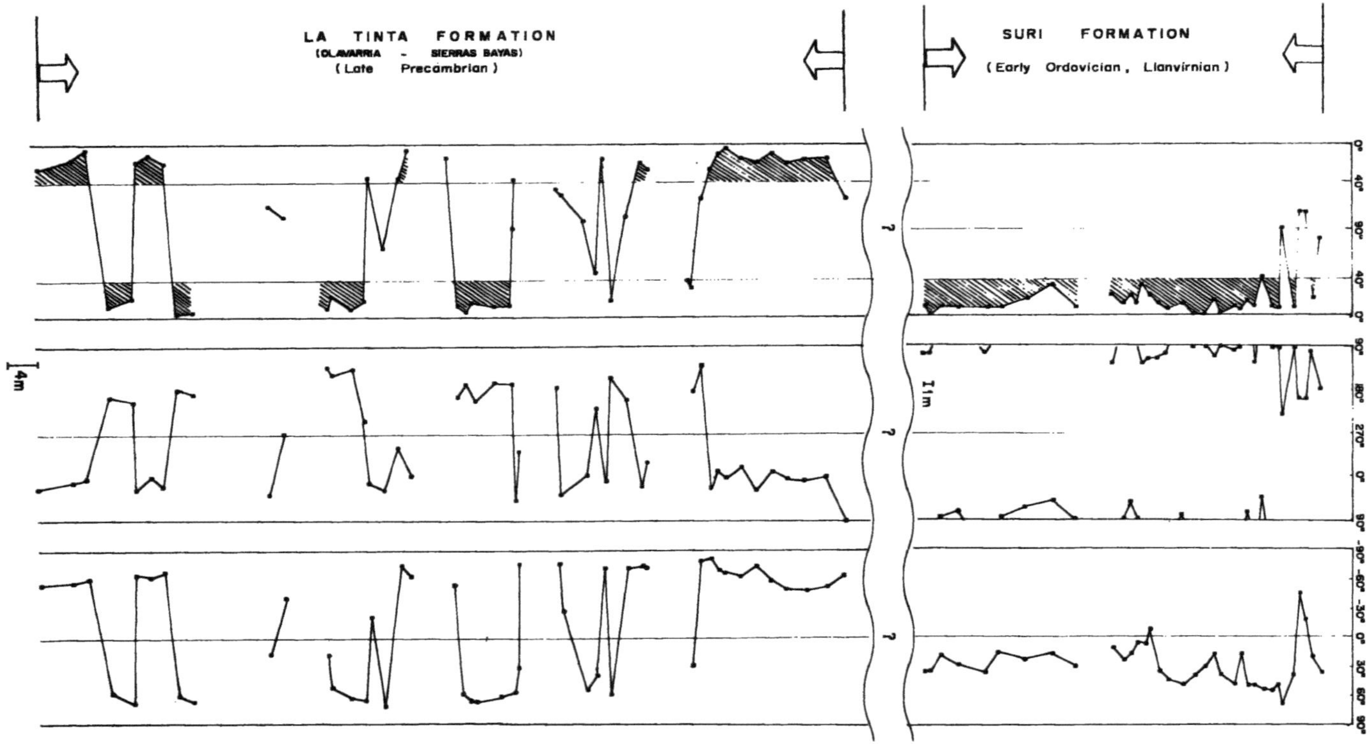

Fig. 1. Mean declination and inclination of stable remanence and the angle between the VGP for each stratigraphic unit and the palaeomagnetic pole through the Suri Formation (Early Llanvirnian, older than 416±24 Myr) and La Tinta rocks (Late Precambrian, 709±24 Myr) from Olavarría-Sierras Bayas, Argentina.

reported (Valencio *et al.*, 1980b). These data are used to define the relative positions of South America and Africa in Late Precambrian time (and therefore to evaluate Du Toit and Dalla Salda interpretations) and some characteritics of the geomagnetic field in that period.

2. Palaeomagnetic Data

Tuffs and shales of Llanvirnian age (K-Ar age older than 416 ± 25 Myr), included in the Suri Formation, are exposed in northwestern Argentina. Eighty-seven hand samples were collected in a sequence of Suri rocks exposed near Puesto Chaschuil (27°47′S, 68°06′W). After AC and thermal cleaning the majority of samples show reversed polarity;

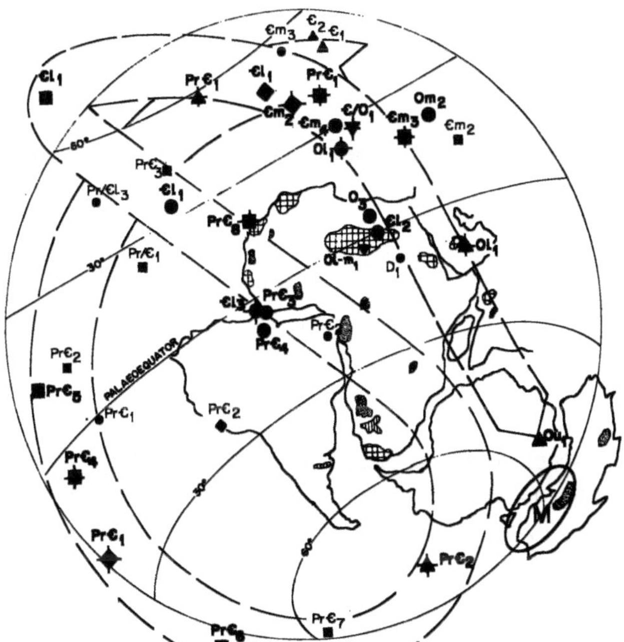

Fig. 2. Pole positions for the gondwanic continents plotted on Vilas and Valencio (1978) reconstruction and joint apparent polar wandering path for the Gondwana for the interval 1,000–400 Myr. Pr€$_3$ (Lower Arumbera and Pertataka Formation), Pr€$_4$ (Upper Arumbera Sandstone) and €l$_3$ (Todd River Dolomite, and Allua and Eninta Formations) are the poles for precambrian and cambrian rocks from Australia reported recently by Kirschvink (1978). ▲, South America; ■, Africa; ●, Australia; ◆, India; ▼, Antartica. ✦, six points; ▲, five points; ▲, four or three points. Lambert's equal area projection; latitude grid based on the magnetism of La Tinta rocks (Late Precambrian). M is the locus of possible pole position for the Marinoan Glacial Beds (about 700 Myr, McElhinny and Embleton, 1976). ▥, Precambrian marine sediments for La Tinta (South America) and Nama Groups (Africa). ▒, Late Precambrian glaciations; ▦, Late Ordovician glaciations after McElhinny and Embleton (1976).

they also record reversals of declination and inclination in the top of the sequence which have been interpreted as geomagnetic excursions (Fig. 1; VALENCIO *et al.*, 1980a). The stable remanence of Suri rocks yield a palaeomagnetic pole at 5.9°E, 8.5°S ($SAOl_1$, $\alpha_{95} = 5.9°$). Following the classifications system for palaeomagnetic data suggested by VILAS and VALENCIO (1978), we have allocated six points (key pole) to $SAOl_1$ (Fig. 2).

Marine sediments of Late Precambrian age, included in La Tinta Formation, are exposed in central Argentina (37.3°S, 59.8°W). Seventy-seven hand samples were collected in sequences of La Tinta rocks exposed near Olavarría-Sierras Bayas and Barker (about 115 km apart). After thermal cleaning and chemical leaching samples show reversed and normal polarity. Figure 1 shows the magnetostratigraphy for La Tinta rocks exposed close Olavarría-Sierras Bayas (709 ± 24 Myr; VALENCIO *et al.*, 1980b). The stable remanence of La Tinta rocks yield a palaeomagnetic pole at 300.9°E, 79.3°S ($SAPr€_2$, $\alpha_{95} = 4.8°$). We allocated six points to $SAPr€_2$ (Fig. 2).

3. Discussion

The palaeomagnetic data for the sequences of Lower Ordovician and Upper Precambrian rocks from Argentina allow us to know some of the characteristics of the geomagnetic field in those periods.

Reversed stable remanence was isolated in Lower Ordovician (Llanvirnian) rocks from Argentina (Fig. 1). This is consistent with the reversed remanence found in Lower Llanvirnian rocks from USSR (North Baikal Superzone; KHRAMOV *et al.*, 1978). Geomagnetic excursions recorded in the top of the sequence of Suri rocks could be related with the starting of the period of normal polarity found in Upper Llanvirnian rocks from USSR.

Several reversals of the geomagnetic field were recorded in Upper Precambrian rocks from Argentina; they define, at least, four periods of normal and four periods of reversed polarity of the geomagnetic field (Fig. 1).

MCELHINNY and EMBLETON (1976) showed that Gondwana was a unit at least as far back as 750 Myr ago; VILAS and VALENCIO (1978) extended this as far as 1,000 Myr ago. Therefore, the comparison of the palaeomagnetic pole positions for La Tinta ($SAPr€_2$, 709 ± 24 Myr) and Suri ($SAOl_1$, Early Llanvirnian) Formations with the Upper Precambrian and Ordovician pole positions for the other gondwanic continents, will allow us to evaluate if they are consistent with the existence of the Gondwana Supercontinent in those periods. To do that we rotated all the reliable palaeomagnetic poles for the gondwanic continents for the Late Precambrian-Early Palaeozoic period to conform with the VILAS and VALENCIO (1978) reconstruction (Fig. 2). This figure shows that the position of $SAOl_1$ is on relatively good agreement with the positions of poles for the Sør Rondane intrusives ($AnOl\text{-}m_1$, Early-Middle Ordovician, 485 ± 25 Myr) for Antarctica and for the Tumblagooda sandstone (AuO_3, Ordovician) and Stairway sandstone ($AuOm_2$, Middle Ordovician) for Australia (references quoted in VILAS and VALENCIO, 1978). The examination of the figure also shows that the position of $SAPr€_2$ is on good agreement with the positions of poles for the Mbozi Complex ($AfPr€_6$, 743 ± 30 Myr) and the Pre Nama dykes ($AfPr€_7$, 653 ± 70 Myr) for Africa and with the locus (M) of the possible pole position for the Marinoan glacial beds (about 700 Myr), for Australia (McElhinny and Embleton, 1976). Therefore, the positions of $SAOl_1$ and $SAPr€_2$ support that Gondwana was a unit in Early Ordovician and Late Precambrian times. This is consistent with the interpretation which suggests that the sediments of Late Precambrian age included in La Tinta (South America) and Nama Groups (Africa, Fig. 2) were deposited in a common shallow-marine basin. However, note that the palaeomagnetic data for the gondwanic continents do not exclude models involving the opening and closing of minor seas.

Finally, the examination of Fig. 2 shows that the joint apparent polar wandering path for Gondwana is consistent with the widespread occurrences of Upper Precambrian and Upper Ordovician glacial deposits across this Supercontinent. This suggests that the timing and location of these deposits are related to a rapid movement of Gondwana over the south pole.

The author wishes to thank the Consejo Nacional de Investigaciones Científicas y Técnicas and the Universidad de Buenos Aires for the support which enabled the work described to be carried on.

REFERENCES

Dalla Salda, L., Nama and La Tinta Groups—a common Southern Africa–Argentine basin?, 16th Annual Report Chamber of Mines, Precambrian Research Unit, Univ. of Cape Town, 1979 (in press).

Du Toit, A., A geological comparison of South America with South Africa with paleontological contribution, F. R. Cowper Reed, Wash. Pub. Carneg. Inst., 381, 1927.

Kirschvink, J. L., The Precambrian-Cambrian boundary problem: paleomagnetic directions from the Amadeus Basin, Central Australia, *Earth Planet. Sci. Lett.*, **40**, 91–100, 1978.

Khramov, A. N., D. M. Pechersky, E. A. Molostovsky, and V. P. Rodionov, Palaeomagnetic scale for the Mesozoic and Palaeozoic of the USSR, Inter-Union Commission on Geodynamics, Workshop on Magnetostratigraphy and Geodynamics, Leningrad, 1978, ICG Scientific Report No. 56, 1978 (in press).

McElhinny, M. W. and B. J. J. Embleton, Precambrian and Early Palaeozoic palaeomagnetism in Australia, *Phil. Trans. R. Soc. Lond.*, **A280**, 417–431, 1976.

Valencio, D. A., J. F. Vilas, and J. E. Mendía, Palaeomagnetism and K-Ar ages of Lower Ordovician and Upper Silurian-Lower Devonian rocks from northwestern Argentina, *Geophys. J. R. Astr. Soc.*, **62**, 1, 27–40, 1980.

Valencio, D. A., A. M. Sinito, and J. F. Vilas, Palaeomagnetism of Late Precambrian rocks of La Tinta Formation, Argentina, *Geophys. J. R. Astr. Soc.*, **62**, 3, 563–576, 1980.

Vilas, J. F. and D. A. Valencio, Palaeomagnetism and K-Ar age of the Upper Ordovician Alcaparrosa Formation, Argentina, *Geophys. J. R. Astr. Soc.*, **55**, 143–154, 1978.

Magnetism of the Mid-Ordovician Tramore Volcanics, SE Ireland, and the Question of a Wide Proto-Atlantic Ocean

Ernst R. Deutsch

Geomagnetic Research Laboratory, Physics Department,
Memorial University of Newfoundland,
St. John's, Newfoundland, Canada

(Received June 25, 1980)

Useful evidence regarding a Proto-Atlantic (Iapetus) ocean may be obtained from palaeomagnetic comparisons across sutures in presumed hybrid blocks such as Newfoundland, Britain and Ireland. Usually this involves Lower Palaeozoic results that may be further compared with data from larger cratonic units, e.g. Europe and North America. Towards these ends the first palaeomagnetic results are reported for an igneous suite comprising andesite sills and flows, pillow lavas, tuffs, mafic intrusives and rhyolite collected from a 20-km wide coastal section near Tramore, Co. Waterford, Eire. From associated fossil evidence the rocks range in age from Lower Caradoc to somewhat earlier (Llanvirn?) Ordovician time, or a mean age of about 460 Myr.

Although at 12 out of the 28 sites studied the remanence of the samples after demagnetization was too weak to measure, or was randomly or anomalously directed, the mean vectors of the remaining 16 sites (76 samples) after AF or thermal treatment and geological tilt correction became well grouped in sharp misalignment with the present Earth's field direction. At most sites the vector groupings of the samples improved upon demagnetization. These results indicate long-term stability of the remanence. At the 6 oldest stable sites this remanence is of reversed polarity and is nearly antiparallel to the (normal) mean direction of the 10 youngest sites, giving an overall north pole at 11°S, 162°W (*dp*, *dm*$=10°$, $13°$; $N=16$ sites). Confidence that the magnetization reflects a mid-Ordovician field direction is based on 1) a strongly positive fold test demonstrating that this magnetization pre-dates the Caledonian folding, 2) the presence of both polarities, and 3) the other long-term stability evidence cited.

The Tramore pole significantly diverges from most published mid- to late Ordovician palaeomagnetic poles for the British Isles, including the Mweelrea ignimbrites in western Ireland. The data support a 30° rotation of the British Isles relative to the geomagnetic field during the Ordovician. The 30° angular difference between the Tramore and Mweelrea poles can be reconciled only through relative rotation between the two sampling localities in a sense that places them on opposite margins of a mid-Ordovician Proto-Atlantic ocean. Application of these Irish palaeomagnetic results to the plate tectonic model of Phillips, Stillman and Murphy yields a unique rotation pivot for post mid-Ordovician closure of a Proto-Atlantic ocean $3{,}300 \pm 2{,}200$ km wide. Ordovician palaeomagnetic comparisons across the North Atlantic suggest that a total Proto-Atlantic reconstruction is premature.

1. Introduction

Three islands, Britain, Ireland and Newfoundland, play key roles in the proposal of a Proto-Atlantic ocean (WILSON, 1966) where each island is considered to be a hybrid block welded from crustal segments that lay on opposite coasts in Lower Palaeozoic times. A palaeomagnetic scheme for testing this proposal (DEUTSCH, 1969) requires that, on any one island, synchronous rocks exposed on opposite sides of the presumed suture will show stable but divergent directions of magnetization. In Ireland, the first pre-Tertiary formations investigated magnetically are the Mweelrea ignimbrites of Llanvirn age, where three studies (DEUTSCH and SOMAYAJULU, 1970; MURTHY and DEUTSCH, 1971; MORRIS *et al.*, 1973) have established a reliable direction of magnetization. The mid-Ordovician age of these rocks is relevant since geological evidence from the Caledonides of the British Isles (DEWEY, 1969) suggests that the ignimbrites are intermediate in age between the (Lower Ordovician) time of maximum width of the presumed Proto-Atlantic and that of its closure at or after the end of the Ordovician.

The Mweelrea ignimbrites are located northwest of the proposed plate suture (PHILLIPS *et al.*, 1976). Comparisons with palaeomagnetic results from Ireland southeast of the suture (P. MORRIS and ROBINSON, 1971; W. A. MORRIS, 1976) have so far proved inconclusive because of discordant magnetization directions among the latter results which the authors variously attributed to remagnetization, insufficient sampling or unknown causes. A relevant though less direct palaeomagnetic comparison involving Ireland is with Britain, where several reliable mid- to late Ordovician results are available (summarized in IRVING *et al.*, 1976; FALLER *et al.*, 1977; PIPER, 1979). Most of the British sites are south of the presumed plate suture and yield in general very similar pole positions from which, moreover, one pole located north of the "suture" (Aberdeenshire gabbros; SALLOMY and PIPER, 1973) departs only marginally. The combined British pole, however, is 30 degrees away from the Mweelrea ignimbrite pole. BRIDEN *et al.* (1973) and MORRIS (1976) concluded that, within palaeomagnetic error, the pole during the mid- to late Ordovician remained essentially stationary relative to the whole British Isles and any closure across the Caledonides must have been small (1,000 $\pm$ 800 km). They proposed that the discordant magnetization direction of the ignimbrites was caused locally by a 30-degree net horizontal rotation of the South Mayo trough and that this invalidates the original explanation (DEUTSCH, 1969) of the Mweelrea result in terms of a large Proto-Atlantic ocean.

In this paper, palaeomagnetic results are presented for the first time from the widely exposed igneous suite near Tramore, County Waterford, southeast of the suture in Ireland. From fossil dating, the rocks are of mid-Ordovician age comparable to the Mweelrea ignimbrite age. The results, when compared with other palaeomagnetic data for the British Isles, call for re-examination of the question of a wide Proto-Atlantic in this section of the Caledonite belt. By combining the palaeomagnetic evidence from the two sides of the Irish suture with a plate tectonic model (PHILLIPS *et al.*, 1976; PHILLIPS, 1980) it is attempted here to estimate quantitatively the width of the inferred Proto-Atlantic ocean in mid-Ordovician times.

2. Geology and Sampling

A 25-km coastal strip of largely volcanic rocks west of Tramore, County Waterford, forms part of a wider belt of Cambrian to Silurian volcanism in south-eastern Ireland which peaked in the Upper Ordovician. The rocks are largely submarine and are considered (PHILLIPS *et al.*, 1976; STILLMAN, 1978; STILLMAN and WILLIAMS, 1978) to have been erupted at a continental plate margin related to a SE-dipping subduction zone and developing arc-trench system that formed part of a Lower Palaeozoic Proto-Atlantic ocean; such a subduction system was postulated earlier for the Lake District and Wales (FITTON and HUGHES, 1970). STILLMAN *et al.* (1974), SCHIENER (1974), and STILLMAN (1978) have described the geology west of Tramore, where the volcanism is believed to have begun in Llandeilo time or somewhat earlier and was interrupted by deposition of the Tramore limestone, a fossil marker horizon of Llandeilo age (CARLISLE, 1979). The volcanism in this area ended in the Lower Caradoc (CARLISLE, 1979) and the rocks were deformed during the Caledonian orogeny. The volcanic rocks of the Waterford/Tramore region are, in decreasing order of abundance, rhyolite, basalt, andesite and dacite, the acidic rocks being predominantly pyroclastic or intrusive (STILLMAN and WILLIAMS, 1978). Alteration seems to be due mainly to weathering during or shortly after emplacement, or hydrothermal metasomism; rocks affected by a superimposed burial metamorphism commonly are of lower than greenshist facies grade (STILLMAN and WILLIAMS, 1978).

Palaeomagnetic collection were made in 1971 and 1976, by the author with logistical assistance from the Geological Survey of Ireland and the Dublin Institute of Advanced

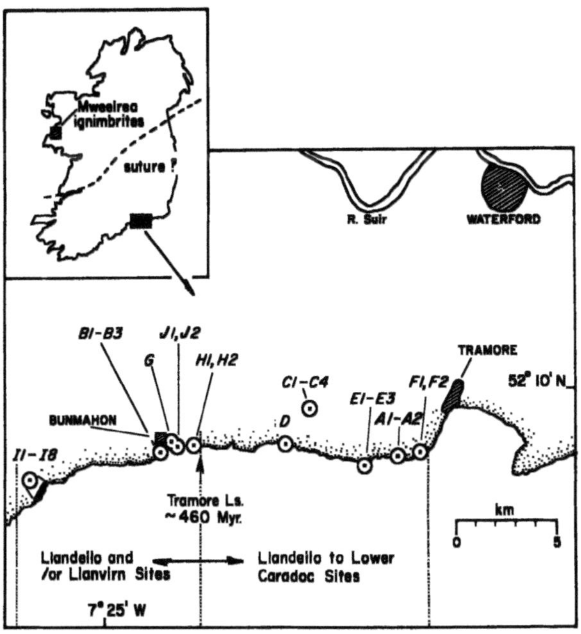

Fig. 1. Tramore area, location of sampling sites.

Table 1. Tramore area: site-mean remanence directions and Fisher statistics.[1]

Site No.[2]	Rock type	Natural remanence (NRM)				Treatment[4]	After demagnetization				After tilt correction		
		D (°)	I (°)	N	k^3		D (°)	I (°)	N	k^3	D (°)	I (°)	α_{95} (°)
A1	Mafic dike or sill	240	0	5	2.1*	40 mT	206	−39	5	4.1	305	−62	42
B1	Andesite (flow?)	307	−15	5	21	40 mT	309	−32	5	14	308	−60	21
B2	As B1	314	+01	4	16	40 mT	315	−28	4	22	317	−56	20
B3	As B1	359	−10	4	<2*	40 mT	353	−43	4	3.2*	—	—	—
C1–C3	Samples from three acid tuff bands[5]	204	+03	5	7.0	—	—	—	—	—	—	—	—
D	Banded rhyolite	142	−64	3	2.8*	40 mT	152	−53	3	2.2*	—	—	—
E1	Acid-intermediate tuff	313	−16	3	109	40 mT	328	−34	3	107	356	−74	12
E2	As E1	261	+06	3	14	40 mT	309	−24	3	33	301	−66	22
E3	Mafic dike	312	+25	3	2.0*	40 mT	331	−54	3	5.9	331	−54	57
F1	Andesite pillows	219	+20	6	6.5	40 mT	237	−13	6	4.7	(230)	(−22)	(34)[10]
F2	As F1	310	−31	5	6.8	40 mT	319	−43	5	70	333	−64	9.2
G	Andesite pillows	215	+55	6	2.7	500 C[7]	294	+32	6	<2*	—	—	—
H1	Andesite flow	289	0	9	9.3	40 mT	273	−37	9	5.3	264	−53	25
H2	As H1	333	+62	5	5.3	40 mT	326	+60	5	93	(317)	(+43)	(8.0)[10]
I1	Andesite sill[6]	263	+58	6	5.1	200 C	241	+49	6	7.1	(1)	(+25)	(27)[10]
I2	As I1	191	−15	5	2.0*	200 C	200	−16	4[9]	12	167	+85	27
I3	As I1	211	+17	5	2.0*	200 C	177	−16	4[9]	4.4	136	+64	49
I4	As I1	188	+49	6	<2*	200 C[8]	194	+10	6	<2*	—	—	—
I5	As I1	218	+11	6	4.7	200 C	208	−08	5[9]	26	183	+71	15
I6	As I1	191	−20	5	2.3*	300 C	190	−18	5	6.7	178	+60	32
I7	As I1	185	−06	5	15	300 C	189	−26	4[9]	37	183	+52	15
I8	As I1	193	+36	5	3.2	300 C	206	+28	5	3.9	103	+53	38
J1	Andesitic tuff	318	−41	6	32	500 C	322	−46	6	116	342	−41	6.2
J2	Andesite sill[6]	314	−08	6	<2*	500 C	315	−41	6	39	333	−39	12

[1] D=declination, deg., E of N; I=inclination, deg., positive downward; N=number of samples averaged (1 specimen per sample); k=precision parameter; α_{95}=radius of 95% confidence circle.

[2] Two tuff sites, A2 and C4 (4 samples each) are omitted as the samples were too weak to measure.

[3] An asterisk (*) after a k-value denotes that within-site vector grouping is random at 95% probability level (IRVING, 1964, p. 63). Sites with random grouping after demagnetization were excluded from final averaging.

[4] Values are for stepwise AF or thermal treatment (see text).

[5] NRM values shown are based on 5 samples from undifferentiated sites C1, C2, C3 measured on the cryogenic magnetometer at Ohio State University. All ten samples from these sites were too weak to measure on the spinner.

[6] Sills probably formed by syn-depositional intrusion into unconsolidated host rock (see text).

[7] Directions are random also at 300 and 600 C.

[8] Directions are random also at 300 and 500 C.

[9] One greatly divergent sample (>65° from mean) was omitted from averaging.

[10] Tilt-corrected sites I1 and F1 were omitted as their directions diverge sharply (>65°) from the respective mean directions of site groupings (D, I=140°, +76° for 7 I-sites incl. I1; D, I=309°, −60° for 11 other sites incl. F1). Discordant site H2 was omitted because of uncertainty in geologic structure (W. Stillman, personal communication).

Studies and with guidance in the selection of sampling sites from Dr. W. Stillman of Trinity College and Dr. E. Schiener of University College, Dublin. A total of 134 oriented drill cores and hand samples were collected from 28 sites (Fig. 1), comprising andesite sills and flows, pillow lava, tuffs, mafic intrusives and some rhyolite (Table 1). The 12 easterly sites (A1, A2, C1–4, D, E1–3, F1, F2) overlie the Tramore limestone and are presumably Lower Caradoc to Llandeilo in age, or about 445–462 Myr on the Elsevier time scale (VAN EYSINGA, 1975). The 16 westerly sites (B1–3, G, H1, H2, I1–8, J1, J2) lie below the limestone and are thus Llandeilo or older (>450 Myr), the oldest sites (I1–8) being probably Llanvirn (M. Boland, personal communication), i.e. 473–462 Myr. Thus the widest probable age range at the 28 sites is 473–445 Myr, with a mean value of about 460 Myr.

The samples were oriented with a Brunton compass which was checked against a solar compass whenever possible, i.e. at most sites. No systematic differences between magnetic and solar readings were found and there seems to be no reason to suspect the Brunton readings taken on other outcrops on sun-less days. Dip and strike values accurate to 2° or better were obtained at nearly every site, usually as the mean of several determinations. At sites other than the I-sites, the dips are typically 20–45°. As there were no indicators, such as plunging fold axes, of serious secondary deformation at any site, it was considered safe to reconstruct the original bedding positions by back-tilting about the strike.

Sites I1 to I8 comprise eight andesite sills, each 10–60 m thick, separated by thick tuff, slate or shale layers. The strata have been tilted by about 90° into their present near-vertical attitudes. These andesites, and probably those at site J2, are believed to have been injected horizontally into the unconsolidated wet sediment close to the sediment-water interface (SCHIENER, 1974). Since sites I1 and I8 (top and bottom sills) are separated by more than 600 m, it is plausible that geologically significant time intervals separated the injection of the different sills and therefore, that secular variation components in the primary remanent magnetization of the sills tend to average out between different sills. The time span represented by the whole Tramore sampling area is almost certainly adequate to have cancelled Earth's field asymmetries of 10^3–10^5 years' duration.

3. Remanent Magnetization

At the laboratory the samples were cut into two or more cylindrical specimens of 2.2 cm diameter and 1.8 cm mean height. The following apparatus was used: (i, ii) Two spinner magnetometers (Princeton Applied Research and Schonstedt); (iii) Alternating-field (AF) demagnetizer with 3-axis tumbler (PEARCE, 1967), for ≦100 mT peak field; (iv) Thermal demagnetizer attaining 800 °C; the original unit (DEUTSCH and SOMAYAJULU, 1970) was improved by addition of large Parry coils and a feedback system confining the estimated maximum field at any specimen to 5 nT. Supplementary equipment included (v) a high-field (≦350 mT peak) AF demagnetizer; (vi) a DC electromagnet for producing high-field isothermal remanence (IRM); and (vii) a bridge for measuring temperature dependence of susceptibility (PÄTZOLD, 1972) between −196 C and 800 C in a field of 0.03 mT. All measurements were made in air.

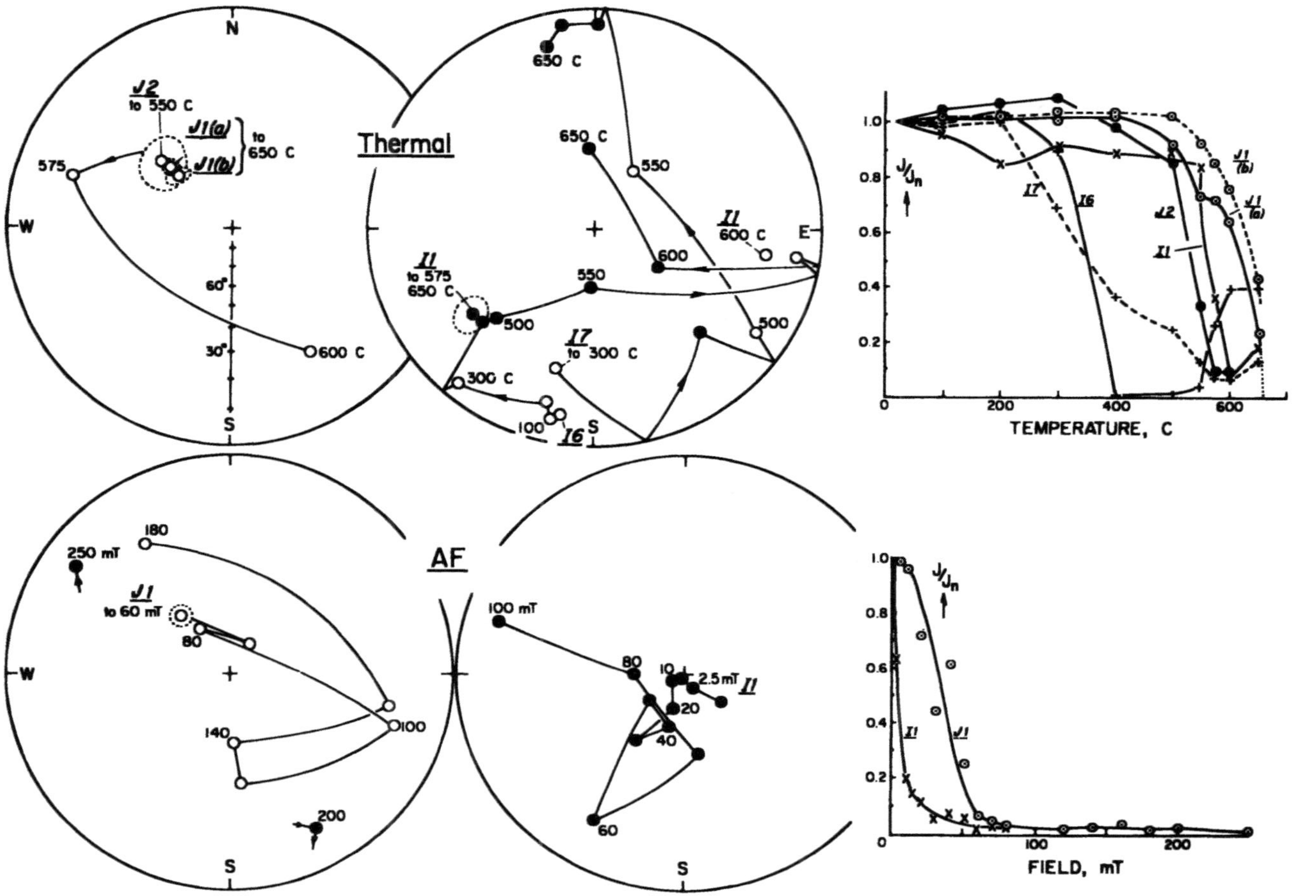

Fig. 2. Pilot thermal and alternating-field (AF) demagnetization of selected andesitic specimens from the sites shown (underlined). Specimens used in comparing thermal and AF treatments were taken from different samples at a given site (J1, I1); also specimens J1(a) and J1(b) are from different samples. Open circles indicate N. pole directions upward; filled circles, downward. Within dashed outlines, directions remain stable up to the highest step shown. The stereograms are Wulff nets.

3.1 Natural remanence (NRM) and related properties

Most of the sampling sites (Fig. 1) are weakly magnetic, though at some sites the NRM intensity (J_n) varies widely. J_n values for 60 samples range from 0.1–0.6 ($\times 10^{-3}$ A/m) for the acid tuffs (sites C1–C3) to 2,000 ($\times 10^{-3}$ A/m) for the exceptionally magnetic "andesitic tuffs" (site J1), and at 17 sites of andesites (sills, pillows, flows) the J_n distribution is 0.2–50 ($\times 10^{-3}$ A/m). The intensities of the rhyolites (site D), mafic intrusives (sites A1, E3) and "acid-intermediate tuffs" (sites E1, E2) also span this broad range. Such variable but relatively low intensities are typical in rocks of this age and composition, making it cumbersome to measure the weakest remanences accurately. Five tuff sites (A2, C1–C4) are omitted from Table 1 due to this cause, but the NRMs of five specimens from sites C1, C2 and C3, treated as one site, were measured on the cryogenic magnetometer at Ohio State University through the courtesy of Prof. H. Noltimier (Table 1).

Initial volume susceptibility (K) was measured on 31 samples (one specimen per sample) of andesite sills (sites I1–I8, J2), pillows (site G) and tuff (site J1). All but one specimen from the 10 sill and pillow sites gave low susceptibilities in the range 0.1–1.1 ($\times 10^{-3}$ SI units), while the "magnetic" site J1 gave values of 2.5–50 ($\times 10^{-3}$ SI units). Koenigsberger ratios $Q_n (= J_n/KH$, where $H = 40$ A/m is the present field) were calculated for these 31 specimens. Only one Q_n value exceeds unity (2.3 at site G), the other 30 being in the range 0.01 to 1. Such findings are not uncommon in older, e.g. Palaeozoic, volcanic rocks and indicate that induced and low-stability components dominate the NRM. While this is partially confirmed in Fig. 2, the low Q_n ratios have not precluded the extraction of a stable remanence through cleaning, usually at intermediate fields or temperatures (Figs. 2 and 5; Table 1). Conversely, the presence of a stable remanence and relatively high Q_n in

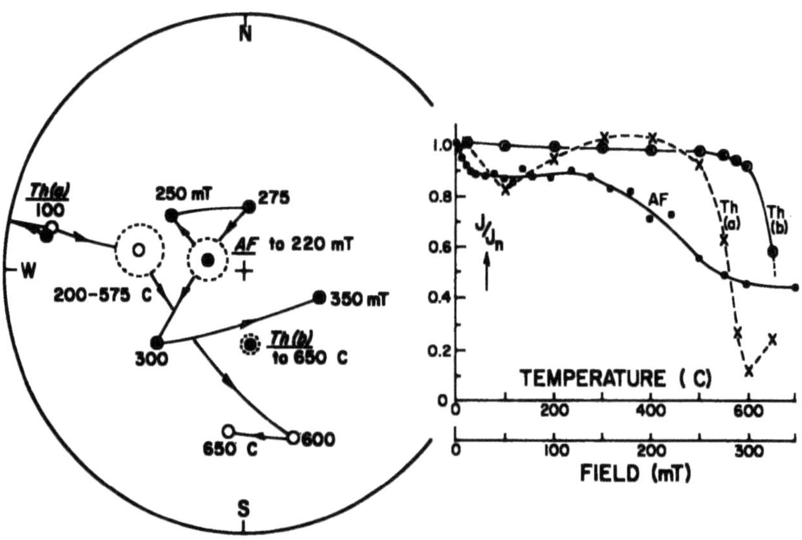

Fig. 3. Pilot thermal and AF demagnetization of selected pillow andesite specimens from the "anomalous" site G. The three specimens are from different samples. Notations are in Fig. 2. Note how sharply these stable remanence directions of different specimens diverge.

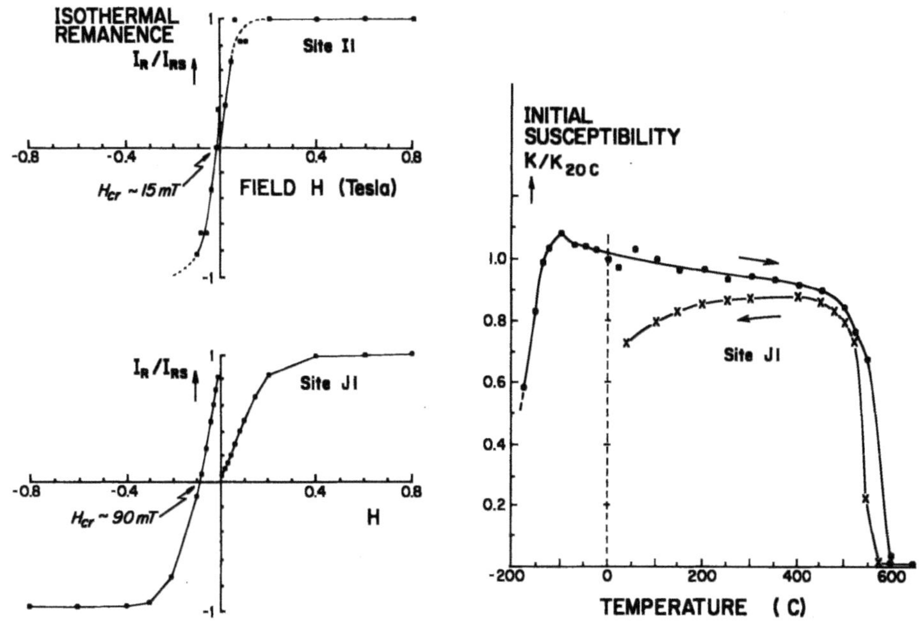

Fig. 4. Other magnetic properties, for three specimens from sites I1, J1. *Left*. The normalized isothermal remanence is shown, where I_R is the remanence acquired stepwise in fields H, I_{RS} the saturation remanence and H_{cr} the coercivity of remanence. The I1 specimen is typically soft, the J1 specimen typically hard. *Right*. Temperature dependence of magnetic susceptibility (K) in low field (0.03 mT), normalized, for a specimen heated and cooled in air. The heating curve indicates multi-domain magnetite (see text), which may depart from stoichiometry since the peak occurs at a temperature above the isotropic point (−155 C) for magnetite.

a suite of rocks does not ensure that their directions are significantly grouped, a case in point being site G (Fig. 3).

Nine of the 24 site-mean NRM directions in Table 1 are random at the 95% confidence level (IRVING, 1964, p. 63), and the precision (k) of the others is generally not high. Figure 5 (upper left) shows that the between-site grouping also is poor.

3.2 Pilot demagnetizations

One specimen each of 15 samples from 11 sites (B1–3, D, E1–3, F1, F2, H1, H2) were given AF treatment in 12 steps to 80 mT. The early steps showed a trend towards steeper negative remanence directions to northwest, compared with the more horizontal NRMs. This trend can be seen best in the site-mean directions (Fig. 5, top). The smallest vector movements and tightest between-sample groupings generally occurred in the AF range 15 to 40 mT, while at higher fields the directions became increasingly randomized. Between the NRM and the 40 mT step the intensity J_n decayed by an order of magnitude in some specimens, but less in magnetically harder specimens.

One specimen each of 29 samples from 11 other sites (I1–8, G, J1, J2) were thermally demagnetized in 9 steps to 650 C. As was found for the AF-treated sites, the demagnetized remanence vectors tend to point upward and to the northwest in the case of

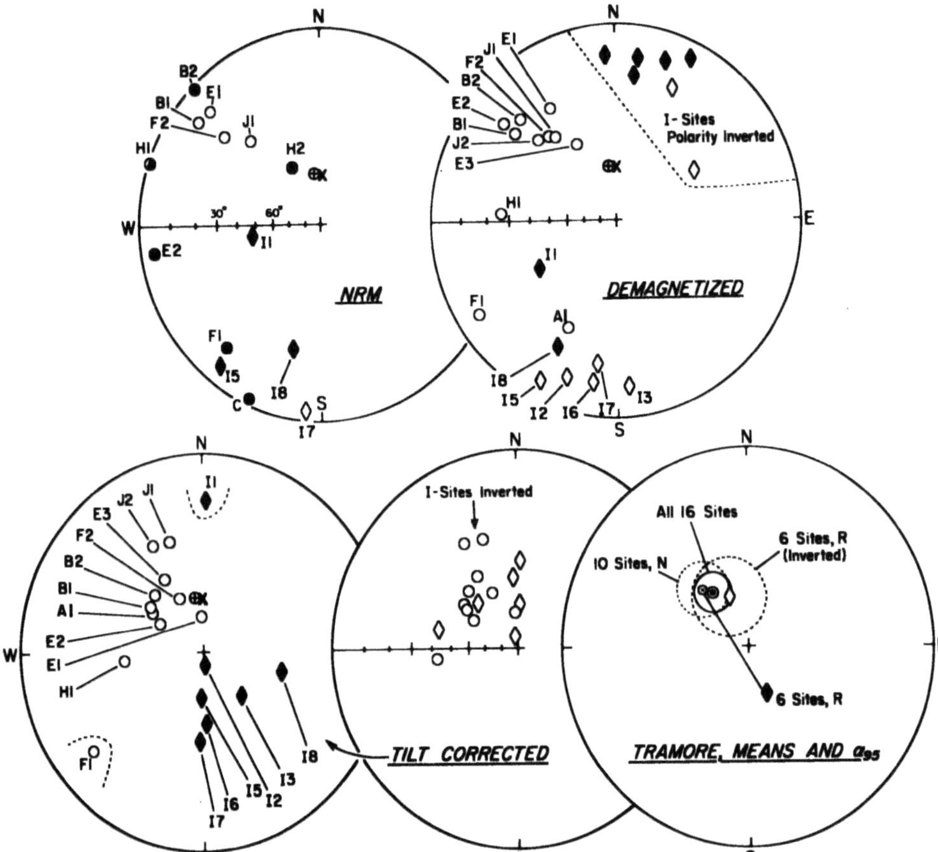

Fig. 5. Site-mean and (bottom right) overall mean remanence directions, excluding sites having random within-site grouping. See also Tables 1, 2. The NRM directions, demagnetized directions (AF or thermal) and demagnetized plus tilt-corrected directions are shown. Diamonds denote I-sites; circles, all other sites. Open symbols indicate N. pole directions upward; filled symbols, downward. X and circled cross denote the axial dipole and present field directions (downward) at Tramore. Dashed lines separating inverted I-sites (top right) and two aberrant sites (bottom left) have no statistical significance. *Bottom right.* Circled dot and open diamond denote group mean directions for the normal (N) and inverted reversed (R) sites, with 95% confidence (α_{95}) circles shown dashed. The N and R mean directions have been joined to depict the slight antiparallelism. The circled thick dot is the Tramore mean direction, shown with solid α_{95} circle. Wulff net.

sites J1 and J2, but for sites I2 through I8 the directions are shallow and southerly (Fig. 5, top), for a reason that becomes clear later. Smallest vector movements and best groupings occurred usually below 300 C, and some directions began to be scattered just above 300 C. Six examples (Fig. 2, top) from moderately to highly stable specimens illustrate these trends. With one exception (I7) the thermal decay curves show mainly single discrete blocking ranges near 400, 600 or 650 C, and these may correspond to a primary component. The reason for the intensity rise at high temperature in three of the curves is not clear, but similar observations have been reported from thermal demagnetization by other workers and may be due to viscous remanence acquisition. Figure 2 (top, middle) suggests that systematic high-temperature components are not the cause.

Six fresh specimens from sites I1, I2, I6, G and J1 were demagnetized in alternating fields between 100 and 350 mT maximum to compare with 13 thermally demagnetized specimens from the same sites. Examples (Fig. 2) show that agreement between the two treatments is excellent in the case of the very stable J1 specimens. The agreement is only moderate for the specimens from site I1 used in the figure, which are representative of less stable behavior at the nine andesite sill sites (compare AF and thermal decay curves for site I1). In the case of site G, three examples (Fig. 3) illustrate the peculiar results produced by both treatments: whereas the remanence is, on the whole, very hard, the directions observed in different specimens are grossly discordant, irrespective of the treatment used. Not surprisingly, site G did not yield a meaningful remanence (Table 1). At the other sites being compared (I1, I2, I6, J1), the thermal treatment tended to produce more consistent results than the AF treatment, while the AF results in turn were more erratic than those of the AF pilot demagnetizations on sites B1–3, E1–3, F1, F2, H1 and H2. Ultimately, thermal demagnetization was found preferable at sites I1–I8, G, J1 and J2, and AF treatment at the other sites.

3.3 General treatment

Accordingly, the 23 sites (omitting C1–C3) were treated as follows: one specimen from every sample was demagnetized by AF (12 sites) or thermally (11 sites) in two to four steps. Table 1 shows the treatments adopted, based on the criterion of best vector grouping.

Four of the 23 demagnetized sites (I2, I3, I5, I7) had initially very poor precision due to the presence of one grossly discordant specimen each, and these 4 specimens were omitted (Table 1) to improve precision, a step that is here justified by the fact that it left the overall mean direction for these four sites unchanged. At 4 other sites (B3, D, G, I4) the within-site vector groupings after demagnetization are random, and these sites were excluded from the final statistical treatment. One high-precision site (H2) was excluded as well, due to uncertainty in the geologic structure at that site, which was not realized until this work was completed (W. Stillman, personal communication). This leaves 18 significant site-mean directions, 14 of them with improved precision due to cleaning.

Demagnetization has resulted in two clearly distinct site groupings (Fig. 5, top right) having remanence directions to the south (mainly I-sites) and northwest, each group showing increased between-site precision compared to the NRM. It was suspected that the two groups might be of opposite polarity, but this cannot be the sole cause of the misalignment which is roughly 90°. The I-sites have been plotted also with inverted polarity to illustrate this fact.

3.4 Remanence stability and origin

Demagnetization has typically removed a component whose mean direction is between the NRM and the present (downward) field direction. This applies to sites in both of the directional groupings and can be verified by inspection of Fig. 5 (top). Exceptions are highly stable specimens (e.g. Fig. 2, left diagrams) that seem to lack a secondary component. At all sites except H2, the site-mean directions both before and after demagnetization depart from the present and axial dipole directions by large angles (Fig. 5, top). These facts, along with the pilot treatment results, support the supposition that the remanence isolated by demagnetization has long-term stability and represents a geologically ancient Earth's field direction.

In the andesite sills (sites I1–8, J2) magnetite or titanomagnetite is probably the only remanence carrier, as is suggested in Fig. 2 by the generally simple trends of the thermal decay curves showing that the magnetization is blocked below the Curie temperature (T_c) of magnetite, and by the low value (<10 mT) of the median demagnetizing field ($H_{1/2}$) for site Il. Although this particular value reflects an untypically low coercivity (H_c), the more typical finding of medium-hard components ($H_c \sim 10$–20 mT) in the andesite sills and also at most other sites is likewise consistent with (titano)magnetite as the sole remanence carrier. This interpretation is supported by a test of isothermal remanence (IRM) using a fresh specimen from site I1 (Fig. 4). High-field IRM is diagnostic of the entire magnetic fraction and shows that this material is saturated below 100 mT and has a low coercivity of remanence ($H_{cr} \sim 15$ mT).

For the exceptionally stable site J1, the decay curves show a hard component ($H_{1/2} \sim 40$ mT) and a blocking range extending to temperatures above 600 C. Site G shows similar properties. IRM curves (Fig. 4) show that saturation is not reached until 400 mT and that H_{cr} is high (~ 90 mT) suggesting that the magnetic carrier at site J1 (and site G?) may be highly cation-deficient magnetite or perhaps magnetite plus hematite. Finally, a test of susceptibility vs temperature (K-T, Fig. 4) yielded a high T_c (600 C) and a low-temperature peak on heating. Such peaks are diagnostic of multi-domain (MD) magnetite (Radhakrishnamurty and Deutsch, 1974). Therefore this specimen contains more than one component, but since MD magnetite is unlikely to contribute to stable remanence, the remanence is probably carried by cation-deficient single-domain magnetite or by hematite or both.

4. The Mid-Ordovician Field in SE Ireland

After correction for geologic tilt the site-mean directions become further polarized in two nearly antiparallel groupings (Fig. 5, bottom left) comprising respectively the I-sites and all other sites. One site-mean direction in each group (F1, I1) is sharply misaligned with the remaining directions in that group and these two sites were omitted from final averaging to improve precision (Table 1). The overall mean direction (Fig. 5, bottom right) is nearly unaffected by this step. Group mean directions for the 16 remaining sites (10 and 6 sites each) are given in Table 2, showing that the tilt correction has moderately improved the between-site precision for each group. This could indicate that the remanence isolated by demagnetization preceeded tectonic deformation.

4.1 Significance of the site groupings

To obtain the mean vector for all 16 sites the two groups can be combined in four ways. The 168° difference between the two tilt-corrected group-mean directions suggests

Table 2. Tramore area: overall mean remanence directions and Fisher statistics, after demagnetization.[1]

		Before tilt correction					After tilt correction				
	N	D(°)	I(°)	R	k	α_{95}(°)	D (°)	I (°)	R	k	α_{95}(°)
All except I-sites[2]	10	307	−43	8.957	8.6	17	319	−59	9.594	22	11
I-sites only[3]	6	195	−10	5.629	14	19	156	+68	5.718	18	16
All sites (I-sites unchanged)	16	258	−46	9.840	2.4	31	307	−46	4.139	<2	69
All sites (I-sites inverted)[3]	16	337	−27	11.270	3.2	25	*324.3*	*−62.6*	*15.243*	*20*	*8.5*

Tramore: present field direction, 349°, +68°; axial dipole direction: 0°, +69°
Palaeomagnetic N pole position:[4] 11°S, 162°W (dp=10°, dm=13°, N=16 sites)

[1] N=number of sites; R=resultant of N unit vectors; other notations as in Table 1.
[2] Normal polarity.
[3] Reversed polarity indicated in Table 3; hence polarity of I-site directions is inverted for final averaging (last row).
[4] Calculated from tilt-corrected data in last row (*italics*).

Table 3. Tramore area: F-ratio test of alternative directional groupings.

Alternative[1]	Overall mean direction based on		G	N	δ_b^2	δ_w^2	δ_b^2/δ_w^2	$F_{2,28,0.01}$	$F_{2,28,0.05}$
	Tilt correction (all sites)	Inverted polarity (I-sites)							
1	no	no	2	16	9.49	0.202	47	5.45	3.34
2	no	yes	2	16	6.63	0.202	33	5.45	3.34
3	yes	no	2	16	22.3	0.098	227	5.45	3.34
4	yes	yes	2	16	0.138	0.098	1.40	5.45	3.34

[1] See also Table 2. G and N=numbers of site groupings and sites. δ_b^2 and δ_w^2 (in radians^2)=between-group and within-group angular variance. Last two columns=F-ratio, $F_{2(G-1),2(N-G),\alpha}$, at 1% and 5% probability levels.

that the respective sites have opposite polarity and that the site-mean directions of one group should be inverted for calculating an overall mean direction. Here one makes two basic assumptions: the sites were magnetized in fields of respectively normal and reversed polarity, and the field of either polarity was that of a centered axial dipole. However, since all I-sites have been tilted through 90 degrees or so, it is not immediately obvious that dual polarity combined with tectonic deformation is the statistically favoured mechanism. Other alternative are: 1) The stable remanence post-dates the tilting and there was no reversal. 2) Remanence post-dates tilting, and reversal occurred. 3) Remanence pre-dates tilting, and no reversal. Fisher statistics for these and the fourth alternative (remanence pre-dates tilting and reversal occurred) are given in Table 2. As is clear also from Fig. 5, alternative 3 corresponds to the lowest precision, but the k-values of alternatives 1 and 2 also are low.

The 4 alternatives were then submitted to an F-ratio test (WATSON, 1956; LAROCHELLE, 1967). To apply this, one must assume that the variances of the two populations (i.e., site groupings) being compared are approximately equal. This assumption is reasonable since in each of the four cases tested the ratio of appropriate k-values (Table 2, first two lines) is 2 or less. Table 3 shows that, for alternative 4, the I-sites are not distinct from the remaining sites at 5% probability level, whereas in each of the three other cases it is more than 99% probable that the two site groupings are distinct. It was concluded that the most probable explanation of the paleomagnetic results is alternative 4, according to which the stable remanence (i) was acquired prior to deformation of the Tramore volcanics and (ii) carries a record of opposite polarities of the Earth's magnetic field.

4.2 *Fold test*

Conclusion (i) represents a positive fold test with a sharply improved precision k (from 3.2 to 20, Table 2). Both the fold test and the presence of both polarities constitute additional evidence for long-term stability of the remanence. The fold test shows that this remanence pre-dates the Caledonian orogeny and, considering the stability evidence already discussed, it is reasonable to conclude that the remanence was in fact acquired at about the time the rocks were formed in the mid-Ordovician.

4.3 Field reversal and palaeomagnetic pole

The mean direction of the larger group of sites (319°, −59°; Table 2) corresponds to a north pole located near the apparent polar wandering path for stable western Europe and so was taken to be of normal (N) polarity; then the second group (I-sites) are reversed (R). Since the I-sites are older, a field reversal from R to N polarity must have occurred sometime between the respective times these sites and the other westerly sites (Fig. 1) became magnetized, i.e. most probably during the Llanvirn or early Llandovery.

The overall mean direction obtained by combining normal and inverted reversed sites is 324.3°, −62.6° ($\alpha_{95}=8.5°$). Assuming an axial dipole field, this corresponds to a geographical north pole in the present Pacific Ocean, at 11°S, 162°W ($dp=10°$, $dm=13°$, $N=16$ sites).

Since at most sites the within-site vector grouping is comparable or even inferior to the between-site grouping, it is difficult to assess any contribution from secular variation to the size of the error oval, and this will not be attempted. A second possible source of scatter is apparent polar wandering within the time span from the oldest (R) to the youngest (N) sites. This is also difficult to estimate, since the 12° departure from antiparallelism of the R and N group mean directions is not significant and neither is the separation between R and N south poles in Fig. 6 (Table 2). For the same reason, however, the magnitude of any drift of SE Ireland or of polar wandering cannot have been large during the age span of Tramore sites.

5. Comparison with Other Palaeomagnetic Data

The Tramore pole (Fig. 6, T) deviates by 30 degrees from a pole (M) of reversed polarity based on the mid-Ordovician Mweelrea ignimbrites (Morris *et al.*, 1973) which

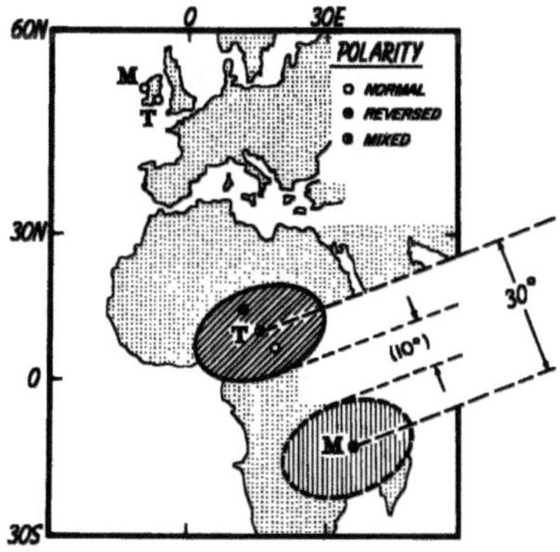

Fig. 6. South pole positions for the Tramore volcanics (T, this study) and Mweelrea ignimbrites (M, Morris *et al.*, 1973) shown with 95% confidence ovals. Pole T is at 11°N., 18°E (*dp*, *dm*=10°, 13°), pole M at 11°S, 38°E (*dp*, *dm*=10°, 18°). Poles are shown separately, without ovals, for the Tramore normal sites (8°N, 21°E; *dp*, *dm*=12°, 16°) and reversed sites (16°N, 13°E; *dp*, *dm*=23°, 28°). Plotted on the present geography.

lie across the proposed plate suture from Tramore. Like the Tramore volcanics, the ignimbrites yield a positive fold test (DEUTSCH and SOMAYAJULU, 1970; MURTHY and DEUTSCH, 1971; MORRIS *et al.*,1973) and the M pole is believed to represent the Llanvirn age of the ignimbrites. As Llanvirn is also the favoured age of the R sites at Tramore, the Mweelrea rocks conceivably became magnetized during the same reversed polarity epoch as the Tramore R sites. Since the R sites predate the N sites, some of which are post-Llanvirn, the Tramore rocks (N and R sites) are *on average* probably somewhat ($\leqq 10$ Myr) younger than the Mweelrea sites.

The M and T pole positions are significantly different and can be reconciled only through relative rotation between the two sampling localities. A notable feature of Fig. 6 is the fact that localities M and T lie approximately on a common great circle with the T poles (reversed, combined, normal) and the M pole, in that order. The great circle is a palaeomeridian and one sees that, *going back in time*, any simple plate rotation designed to make the M and T poles coincide must involve an increasing separation between the M and T sampling localities roughly along the present NW-SE trend (i.e., ancient N-S trend) of the alignment. This trend is nearly perpendicular to the strike of the plate suture (62°: PHILLIPS *et al.*, 1976). Such a reconstruction is consistent with the hypothesis that the two localities, which are now 2° apart, lay on opposite (north and south) sides of a mid-Ordovician Proto-Atlantic ocean. From the palaeolatitudes of Mweelrea ($\lambda_p = 14°$S) and Tramore ($\lambda_p = 44°$S) this ocean at least partly occupied the southern hemisphere and was at least 28° wide, not considering palaeomagnetic error.

I had argued earlier (DEUTSCH, 1969) that a large discrepancy between the then available Mweelrea pole (DEUTSCH and SOMAYAJULU, 1970) and British Ordovician poles was consistent with the existence of a possibly wide Proto-Atlantic ocean in this section of the Caledonides. While further work, (MURTHY and DEUTSCH, 1971; MORRIS *et al.*, 1973) left the Mweelrea pole essentially unchanged, it was counter-argued (BRIDEN *et al.*, 1973; MORRIS, 1976) that the aberrant magnetic direction of the ignimbrites relative to British Ordovician rocks is due to local tectonics in NW Ireland rather than an intervening ocean. Specifically, MORRIS (1976) reasoned that development of the South Mayo trough during the Devonian caused the ignimbrites to rotate first 60° anticlockwise, then 30° clockwise, producing the observed southeasterly (130°–135°) declination; one assumes here that the original direction was parallel to the south or north (~170°, ~350°) declinations characteristic of British Ordovician localities. As a further argument against a substantial Proto-Atlantic, BRIDEN *et al.*, (1973) pointed out that the small difference between the observed magnetization directions north of the hypothetical suture in Britain (Aberdeenshire gabbros: SALLOMY and PIPER, 1973) and south of the suture is not significant.

These were persuasive objections in the light of then available palaeomagnetic data. The results presented here override the objections and require reconsideration of the case for a wide Proto-Atlantic. IN Fig. 7, the results of nine palaeomagnetic studies of British and Irish Ordovician rocks are plotted as south poles. Only those studies are included in which it is reasonably sure that the magnetization was acquired within the Llanvirn-Llandeilo-Caradoc time span (473–440 Myr). Several reliable but anomalous mid- to late Ordovician poles in high latitudes (THOMAS and BRIDEN, 1976; PIPER, 1979) have been

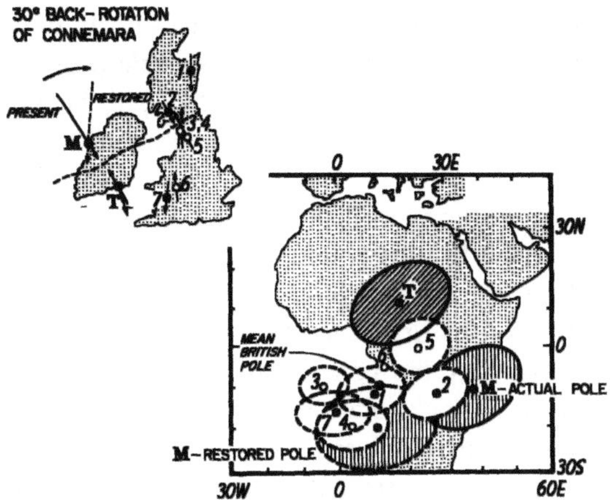

Fig. 7. Palaeomagnetic south poles from mid- to late Ordovician (Llanvirn to Caradoc) igneous rocks in the British Isles, plotted on the present map. The Irish poles (T, M) are as in Fig. 6, pole M being shown also in a "restored" position obtained by reversing a postulated 30° net anticlockwise tectonic rotation of the region bearing locality M. Open, filled and divided circles denote normal, reversed and mixed polarity. Numbers in square brackets below are declinations of the remanence vector shown by arrows on the location map. *British poles*, 1. Aberdeenshire gabbros, Scotland [168°] (SALLOMY and PIPER, 1973). 2. Ballantrae gabbros and serpentinites, Scotland [148°] (PIPER, 1978). 3. Eycott Group lavas, England [0°] (BRIDEN and MORRIS, 1973). 4. Carrock Fell gabbros-granophyres, England [350°] (BRIDEN and MORRIS, 1973). 5. Borrowdale volcanics, England [329°] (FALLER *et al.*, 1977). 6. Breidden Hill intrusive inlier, Wales [341°] (PIPER and STEARN, 1975); for clarity, the large confidence oval (dp, $dm = 15°, 22°$) has been omitted. 7. Builth volcanics, Wales [177°] (PIPER and BRIDEN, 1973). The mean position of poles 1–7 (large filled circle) is 10°S, 10°E. *Irish poles*. M. Mweelrea ignimbrites [131°] (MORRIS *et al.*, 1973). T. Tramore volcanics [N sites, 319°; R sites, 156°; combined sites (R inverted), 324°] (This study). The mid-Ordovician palaeolatitudes of Mweelrea (M) and Tramore (T'') (Fig. 8) are 14°S and 44°S, respectively.

excluded as it seems that these results cannot be explained by regional tectonics (PIPER, 1979) and so are not directly relevant here.

Five of the seven British poles (Fig. 7, nos. 1, 3, 4, 6, 7) were obtained from rocks with average north-south magnetization. When the Mweelrea ignimbrites are rotated 30 degrees clockwise so as to reverse the inferred net deformation of Connemara, pole M becomes statistically coincident with these British poles, apparently conforming to the argument of BRIDEN *et al.* (1973) and MORRIS (1976). However, the local restoration of Connemara not only fails to reduce the discrepancy between poles M and T, but makes it worse. This follows directly from the 30° palaeolatitude difference and the shared palaeolongitude of localities M and T, showing that no amount of horizontal rotation at either locality will reconcile the two poles. Moreover, two poles reported since 1976 (nos. 5, 2) are associated with mean NW-SE declinations and are each distinct from British poles 1, 3, 4 and 7: Pole 5 (Borrowdale volcanics) is an improved result from the English Lake District (previous declination = 0; BRIDEN and MORRIS, 1973) south of the plate suture in Britain, and pole 2 is derived from the Southern Uplands Blocks, north of the suture. These two pole positions are statistically identical to Irish poles T and M, respectively. Then two poles (M and 2) based on rocks north of the suture diverge from the Tramore pole and, marginally, from the Borrowdale pole, both representing rocks south of the suture. As each of these four poles (M, 2, T, 5) corresponds to a NW-SE directed magnetization of similar age, it is difficult to maintain an anomalous status for pole M.

Thus, while tectonic flexuring in NW Ireland cannot be excluded as a cause of the remanent declination of the Mweelrea ignimbrites, the deformation responsible would have to have affected a much larger region of Ireland and Britain on both sides of the suture to explain plausibly why the (NW-SE) declinations at four localities conform. It seems highly improbable that the concordant vector orientations at these localities resulted from separate local block rotations about vertical axes.

It remains to account for the presence of two sets of Ordovician exposures in the British Isles, each set with characteristic declinations (N-S, NW-SE). It seems clear that this requires a change in direction of the Earth's magnetic field relative to the exposures, occurring *during* the Ordovician. Alternative mechanisms could be 1) plate motion implying, however, that *two* plates that were separated by an ocean shared a rotation in the same sense; 2) polar wandering; or 3) an excursion of the geomagnetic field. Further palaeomagnetic work is needed to solve this problem.

6. Width of the Proto-Atlantic Ocean

The postulate of an open Proto-Atlantic ocean in the Ordovician will now be further considered with respect to Ireland only. The palaeomagnetic data have demonstrated the *existence* of such an ocean, but its *size* (other than *minimum* size, from palaeolatitude differences) is more elusive. The ancient separation of a pair of sampling localities now on opposite sides of a suture (e.g. Tramore and Mweelrea) cannot be determined from the two pole positions unless the pivot (rotation pole) is known, which is seldom the case. The pivot position can be determined, however, if the directions of approach of the two plates effecting ocean closure are known or may be assumed.

For this purpose I have taken the Caledonian plate tectonic model of PHILLIPS *et al.*, (1976) as a working hypothesis in testing the palaeomagnetic results. This model represents an attempt to reconstruct in detail the Ordovician to Devonian continental drift history of the British Isles, and it includes quantitative estimates of the geometry of ocean closure. Although Phillips *et al.* estimated also the minimum Proto-Atlantic width (600–800 km in the Arenig) from calculated closure rates, an estimate of the actual ocean width exclusively from their model would involve much extrapolation, which is not the case when palaeomagnetic poles are being compared. The main relevant features of the model are, briefly: a non-parallel approach of two plates ("NW plate" and "SE plate") and their associated subduction systems in the Ordovician, resulting in oblique collision from late Ordovician to early Devonian time and subsequent (Devonian) dextral strike-slip movement by 1,270 km along the ENE-WSW trending suture. The directions of approach relative to the present geography are 155° (NW plate) and 305° (SE plate), allowing for 50% post-collision strain.

I assumed for convenience that the SE plate moved alone relative to a "fixed" NW

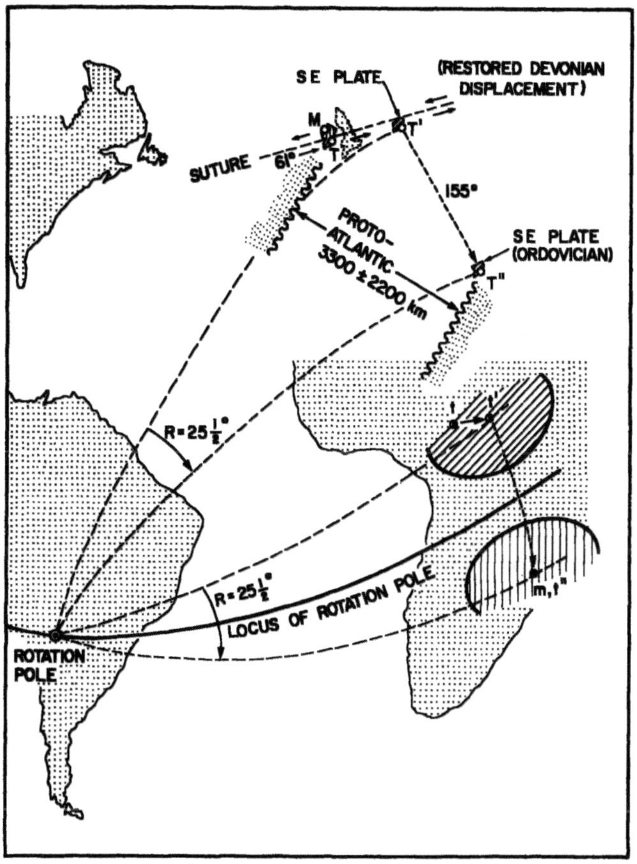

Fig. 8. Proto-Atlantic reconstruction relative to the Irish Caledonides, plotted on the present map (see text). Continental drift patterns inferred by PHILLIPS *et al.* (1976) have been combined with the palaeomagnetic results to yield a rotation pole (27°S, 67½°W) for Proto-Atlantic closure in post-mid-Ordovician time. Poles t, m correspond to the Tramore and Mweelrea poles (T, M) in Fig. 6, and t′ (12°N, 26°E) is the new position of pole t which corresponds to the restored position T′ (56½°N, 11°E) of Tramore when the inferred Devonian strike-slip movement of SE Ireland is reversed. The rotation pole locus is equidistant from poles t′ and m. Rotation, backwards in time, of the SE Plate from T′ to its "final" (i.e. Ordovician) location T″ (34½°N, 27°E) about the rotation pole corresponds to rotation of pole t′ into coincidence with pole m (=t″). The dashed lines from the rotation pole to points T′ and T″ may be visualized as plate lever arms intersecting the drift path T′ T″ at a right angle. The limits (±2,200 km) quoted for the Proto-Atlantic width estimate correspond to the combined 20° palaeomagnetic error in the two confidence ovals.

plate, approaching it in a 320° direction which is the resultant of the separate NW and SE plate motions. For reconstructing the drift geometry one reverses the above sequence (Fig. 8), starting with a 1,270 km eastward displacement of the SE plate along the suture that carries the Tramore volcanics to an "early Devonian" location (T′). Going further back in time, one "opens" the ocean by moving the SE plate in an intial 155° direction from site T′ (obtained by reversing the 320° approach direction and adding 15° for convergence of meridians between sites T and T′). A single-step motion is assumed. To be found is the original (Ordovician) position T″ of the SE plate which results when the displaced Tramore pole t′ is made to coincide with the Mweelrea pole (m). The unknown angular displacement $p(=T'T'')$ is given by (DEUTSCH, 1969):

$$1-\cos p=\sin^2\theta\,(1-\cos R) \qquad (1)$$

$$1-\cos p'=\sin^2\theta'\,(1-\cos R) \qquad (2)$$

where p' is the (known) displacement t′m (=t′t″) between palaeomagnetic poles, θ and θ' are the distances from the rotation pole to T′ and t′ respectively, and R is the angle through which the plate (T′) and the pole (t′) are rotated. Then the rotation pole (or its antipole) lies at one intersection of the rotation pole locus with a great circle that is drawn orthogonally to the drift path T″T′ at point T′. The rotation pole thus obtained (27°S, 67½ °W) determines θ and θ', yielding R (25½°) and p (24½°) in Eqs. (1) and (2).

From p one finds the pre-closure position T″ of the SE plate and so the width TT″ of the inferred Proto-Atlantic ocean relative to the Irish Caledonides. The value obtained (Fig. 8) is 30° ± 20° (or 3,300 ± 2,200 km), where the limits correspond to palaeomagnetic error using the 10° least distance between the two confidence ovals as a rough criterion for the minimum width of the ancient ocean. The 3,300 km value is comparable with the present width of the North Atlantic, showing that the palaeomagnetic results reported here are compatible with a wide Proto-Atlantic ocean in mid-Ordovician time.

After the reconstruction of Fig. 8 had been made, PHILLIPS (1980) proposed a revision of the PHILLIPS *et al.*, (1976) model, in which the postulated angle of approach of the SE plate and magnitude of the subsequent dextral strike-slip movement are changed to 295° and 600 km, respectively. It can be shown, however, that substitution of these figures makes very little difference to the value of Proto-Atlantic width calculated from the older model. In either case a reduction of the large confidence limits associated with that value will be possible when Ordovician palaeomagnetic data having less error become available.

6.1 Comparisons across the North Atlantic

In making the above reconstruction (Fig. 8) the present geographical coordinates have been used as a convenient reference. Allowance for the post-Palaeozoic opening of the Atlantic (BULLARD *et al.*, 1965) must be made, however, if Irish or British Lower Palaeozoic palaeomagnetic results are to be compared with North American results of the same age. Such comparisons are a necessary step towards reconstructing the total Proto-Atlantic palaeogeography which is now poorly understood. For example, the two Irish Ordovician pole positions (Tramore, Mweelrea), after correction for the Bullard fit, still deviate considerably from representative mid- to late Ordovocian poles for cratonic North America (MCELHINNY and OPDYKE, 1973; VAN DER VOO and FRENCH, 1977). Also, early Ordovician poles for eastern and western Newfoundland are far apart (DEUTSCH and RAO, 1977), compatible with a substantial Proto-Atlantic ocean whose size, however, cannot be quantified on the basis of existing Newfoundland data. These problems will be discussed in a subsequent paper.

It is a pleasure to acknowledge the following help received. Logistical support for field trips was generously provided in 1971 by Dr. C. E. Williams, Director, Geological Survey of Ireland and in 1976 by Prof. Thomas Murphy, Director, Dublin Institute of Advanced Studies. Dr. Elmar Schiener, University College, Dublin (1971) and Dr. Chris Stillman Trinity College, Dublin (1971, 1976), gave detailed geological information in the field and later on. David Waddell (G. S. I.) and Dr. David Howard (D.I.A.S.) assisted the author in the field. Maeve Boland, Dr. Adrian Phillips (Trinity College) and others provided crucial discussion. Jim Sharpe and David Morgan did most of the laboratory measurements at Memorial University of Newfoundland. Prof. Hal Noltimier, Ohio State University, made his cryogenic magnetometer available for measuring some weak specimens. This research was supported by Grant A-1946 from the Natural Science and Engineering Research Council of Canada.

REFERENCES

Briden, J. C. and W. A. Morris, Palaeomagnetic studies in the British Caledonides—III. Igneous rocks of the Northern Lake District, England, *Geophys. J.*, **34**, 27–46, 1973.

Briden, J. C., W. A. Morris, and J. D. A. Piper, Palaeomagnetic studies in the British Caledonides—VI. Regional and global implications, *Geophys. J.*, **34**, 107–134, 1973.

Bullard, E. C., J. E. Everett, and A. G. Smith, The fit of the continents around the Atlantic, *Phil. Trans. Roy. Soc. Lond., Ser. A.*, **258**, 41–75, 1965.

Carlisle, H., The Ordovician stratigraphy and faunas of the Tramore area of Co. Waterford with a revised Ordovician correlation for east and south-east Ireland, in *The Caledonides of the British Isles Reviewed*, Spec. Publ. No. 8, edited by A. L. Harris, C. H. Holland, and B. E. Leake, Geol. Soc. Lond., 1979.

Deutsch, E. R., Paleomagnetism and North Atlantic paleogeography, in *North Atlantic—Geology and Continental Drift, Memoir 12*, edited by M. Kay, pp. 931–954, Am. Assoc. Petrol. Geol., Tulsa, 1969.

Deutsch, E. R. and C. Somayajulu, Palaeomagnetism of Ordovician ignimbrites from Killary Harbour, Eire, *Earth Planet. Sci. Lett.*, **7**, 337–345, 1970.

Deutsch, E. R. and K. V. Rao, New palaeomagnetic evidence fails to support rotation of western Newfoundland, *Nature*, **266**, 314–317, 1977.

Dewey, J. F., Evolution of the Appalachian/Caledonian orogen, *Nature*, **222**, 124–129, 1969.

Faller, A. M., J. C. Briden, and W. A. Morris, Palaeomagnetic results from the Borrowdale Volcanic Group, English Lake District, *Geophys. J.*, **48**, 111–121, 1977.

Fitton, J. G. and D. J. Hughes, Volcanism and plate tectonics in the British Ordovician, *Earth Planet. Sci. Lett.*, **8**, 223–228, 1970.

Irving, E., *Palaeomagnetism and its Application to Geological and Geophysical Problems*, Wiley, New York, 1964.

Irving, E., E. Tanczyk, and J. Hastie, *Catalogue of Paleomagnetic Directions and Poles, Third Issue, Paleozoic Results*, Earth Phys. Branch Publ. 75, Dept. Energy Mines and Resources, Ottawa, 1976.

Larochelle, A., A re-examination of certain statistical methods in palaeomagnetism, *Geol. Surv. Canada Paper*, 67–18, 1967.

McElhinny, M. W. and N. D. Opdyke, Remagnetization hypothesis discounted—a paleomagnetic study of the Trenton limestone, New York State, *Geol. Soc. Am. Bull.*, **84**, 3697–3708, 1973.

Morris, P. and K. W. Robinson, Some palaeomagnetic results from the Ordovician andesites of Co. Dublin and Co. Kildare, Ireland, *Geol. Surv. Ireland, Bull.*, **1**, 151–161, 1971.

Morris, W. A., Paleomagnetic results from the Lower Paleozoic of Ireland, *Can. J. Earth Sci.*, **13**, 294–304, 1976.

Morris, W. A., J. C. Briden, J. D. A. Piper, and J. T. Sallomy, Palaeomagnetic studies in the British Caledonides—V. Miscellaneous new data, *Geophys. J.*, **34**, 69–106, 1973.

Murthy, G. S. and E. R. Deutsch, Further paleomagnetic results from the Killary Harbour folded ignimbrites, western Ireland, *Trans. Am. Geophys. Union* (Abstract), **52**, 189–190, 1971.

Pätzold, R. R. High Temperature magnetic susceptibility bridge, M. Sc. thesis, Memorial University of Newfoundland, 1972.

Pearce, G. W., Design and calibration of apparatus for the alternating field demagnetization of rocks, M. Sc.

thesis, Memorial University of Newfoundland, 1967.

PHILLIPS, W. E. A., C. J. STILLMAN, and T. MURPHY, A Caledonian plate tectonic model, *J. Geol. Soc. Lond.*, **132**, 579–609, 1976.

PHILLIPS, W. E. A., Estimation of the rate and amount of absolute lateral shortening in an orogen, using diachronism and strike-slipped segments, Spec. Publ. No. 9, Geol. Soc. Lond., 1980 (in press).

PIPER, J. D. A., Palaeomagnetism and palaeogeography of the Southern Uplands block in Orodovician times, *Scott. J. Geol.*, **14**, 93–107, 1978.

PIPER, J. D. A., Aspects of Caledonian palaeomagnetism and their tectonic implications, *Earth Planet, Sci. Lett.*, **44**, 176–192, 1979.

PIPER, J. D. A. and J. C. BRIDEN, Palaeomagnetic studies in the British Caledonides—I, Igneous rocks of the Builth Wells-Llandrindod Wells Ordovician inlier, Radnorshire, Wales, *Geophys. J.*, **34**, 1–12, 1973.

PIPER, J. D. A. and J. E. F. STEARN, Palaeomagnetism of the Breidden Hill (Palaeozoic) inlier, Welsh Borderlands, *Geophys. J.*, **43**, 1013–1016, 1975.

RADHAKRISHNAMURTY, C. and E. R. DEUTSCH, Magnetic techniques for ascertaining the nature of iron oxide grains in basalts, *J. Geophys.*, **40**, 453–465, 1974.

SALLOMY, J. T. and J. D. A. PIPER, Palaeomagnetic studies in the British Caledonides—II. The Younger Gabbros of Aberdeenshire, Scotland, *Geophys. J.*, **34**, 13–26, 1973.

SCHIENER, E. J., Syndepositional small-scale intrusions in Ordovician pyroclastics, Co. Waterford, Ireland, *J. Geol. Soc. Lond.*, **130**, 157–161, 1974.

STILLMAN, C. J., Day 3: South-east County Waterford and South Tipperary: Ordovician volcanic and sedimentary rocks and Silurian turbidites, in *The Caledonian and Pre-Caledonian Rocks of South-East Ireland, Guide Series 2*, edited by P. M. Brück, P. R. R. Gardiner, M. D. Max, and C. J. Stillman, pp. 41–60, Geol. Surv. Ireland, Dublin, 1978.

STILLMAN, C. J. and C. T. WILLIAMS, Geochemistry and Tectonic setting of some Upper Ordovician volcanic rocks in east and southeast Ireland, *Earth Planet. Sci. Lett.*, **41**, 288–310, 1978.

STILLMAN, C. J., K. DOWNES, and E. J. SCHIENER, Caradocian volcanic activity in east and south-east Ireland, *Sci. Proc. Roy. Dublin Soc.*, **5A**, 87–98, 1974.

THOMAS, C. and J. C. BRIDEN, A geomagnetic anomaly during the late Ordovician, *Nature*, **259**, 380–382, 1976.

VAN DER VOO, R. and R. B. FRENCH, Paleomagnetism of the late Ordovician Juniata formation and the remagnetization hypothesis, *J. Geophys. Res.*, **82**, 5796–5802, 1977.

VAN EYSINGA, F. W. B. (Compiler), *Geological Time Table*, 3rd Edition, Elsevier, Amsterdam, 1975.

WATSON, G. S. Analysis of dispersion on a sphere, *Monthly Notices, Roy. Astron. Soc. Geophys. Suppl.*, **7**, 153–159, 1956.

WILSON, J. T., Did the Atlantic close and then reopen?, *Nature*, **211**, 676–681, 1966.

The Geomagnetic Field during Palaeozoic Time

A. N. KHRAMOV and V. P. RODIONOV

All-Union Oil Research Geological Prospecting Institute, Leningrad, USSR

(Received December 13, 1979; Revised July 10, 1980)

The Palaeozoic reversal sequence is represented by complex rhythms with characteristic times of $(5–7)\times 10^7$, $(1–4)\times 10^7$ and $(1–2)\times 10^6$ y. The first order units recognized are Arginian (R) 590–535 m.y., Siberian (Rn) 535–485 m.y., Baikalian (Nr) 485–430 m.y., Sayanian (Rn) 430–360 m.y., Donetsk (Rn) 360–300 m.y., Kiaman (R) 300–240 m.y., and Illawarra (NR) 240–215 m.y. Three types of Palaeozoic field transitions have been recognized: 1. "sudden" reversals (single or repeated), 2. reversals with a considerable dispersion of palaeomagnetic directions, 3. reversals with loop-wise rotations of the field. The latter may result from the concurrence of axial and equatorial dipoles. Reversals usually occur with a 4 to 10 times lower field intensity. The time span for the complete transition ranges from 8×10^3 to 3×10^5 y. Peculiarities of the reversals have been traced over strike distances of 1600 km. Angular constituents of paleosecular variation have components with characteristic times of $(4–8)\times 10^3$ y. and 3×10^4 y. The latter is characterized by standard deviations of 6 to 10 degrees in stable field epochs and up to 20 degrees in intervals of frequent reversals.

The field of the geocentric axial dipole is a good first approximation of time-averaged field. However, the occasional presence of an irreversible equatorial dipole, controls the shape of geomagnetic reversals of the 3rd type.

1. Introduction

Much effort in palaeomagnetism is focused on problems concerning the areal distribution and changes with time of the geomagnetic field for recent and near past geological epochs (POSPELOVA, 1973; BUCHA, 1973; CREER *et al.*, 1976; NAKAJIMA and KAWAI, 1973).

For the purpose of developing a physical theory of the earth's magnetism it is very important to understand the spatial and temporal characteristics of the geomagnetic field during ancient epochs. The main thrust of Palaeozoic palaeomagnetic studies in the USSR (BUROV *et al.*, 1977; MOLOSTOVSKY *et al.*, 1976; KHRAMOV *et al.*, 1974) is establishing these characteristics.

The present paper deals with evidence on the Palaeozoic palaeomagnetism of the USSR used for scaling geomagnetic reversals in th Palaeozoic and identifying a succession of reversals. Their distribution in time and a study of processes of reversals and secular variation of the geomagnetic field are also described.

2. Palaeomagnetic Scale for the Palaeozoic of the USSR

Early Palaeomagnetic studies for the USSR were aimed towards the solution of specific stratigraphic problems. However these studies provided a great amount of data of variable reliability and completeness. Systematic usage of the data available became possible following palaeomagnetic studies carried out especially to compile standard regional palaeomagnetic scales for some intervals in the Palaeozoic (RODIONOV, 1966; KHRAMOV, 1963a, b). Concurrently first attempts were made to synthesise results for the Palaeozoic segment of the scale (KHRAMOV *et al.*, 1965).

A particular feature of the palaeomagnetic scale in the USSR is the mode of its compilation. It is compiled as a stratigraphic scale on the basis of palaeomagnetic studies of the most complete sections, type locality sections, correlation of sections and construction of standard palaeomagnetic scales for stages, series and systems. This is the main difference between the palaeomagnetic scale for the USSR compared with the widely known scales of Cox, the Lamont scale and other similar scales of geomagnetic reversals which are compiled as time scales based on absolute age determinations (COX, 1969; LARSON and PITMAN, 1972). A magnetostratigraphic method which is preferred for establishing older parts of the palaeomagnetic scale has recently become widely used in other countries too (DACHROTH, 1976; KIRSCHVINK, 1978; PETERSON and NAIRN, 1971). It widens the range of correlation of various palaeomagnetic scales and further confirms the palaeomagnetic scale of the USSR. In the USSR the most representative evidence on the pattern of the palaeomagnetic scale is available for the Cambrian, Ordovician, Upper Palaeozoic and Lower Triassic (BORONIN and BUROV, 1973; DAVYDOV and KRAVCHINSKY, 1970; MOLOSTOVSKY, 1970; RODIONOV, 1966; SLAUTSITAIS, 1975; KHRAMOV *et al.*, 1965).

The publication of reviews on the magnetostratigraphy of the Palaeozoic (MOLOSTOVSKY *et al.*, 1976; KHRAMOV *et al.*, 1974; KHRAMOV, 1975) together with new data provide a more detailed and reliable version of a unified Palaeozoic palaeomagnetic scale for the USSR and can be used as a basis for establishing the history of geomagnetic reversals and global correlations.

Major palaeomagnetic units corresponding to a rank of stratigraphic systems can now be established, e.g. hyperzones (10^7–10^8 y.), for the best defined intervals within an N- or R-hyperzone a subdivision into minor units is also recognized, viz. superzones and zones. These divisions are labelled according to guidelines set down for magnetostratigraphic nomenclature. The description of the palaeomagnetic scale for the Mesozoic and Palaeozoic is presented below (Fig. 1).

The Lower Palaeozoic part (Cambrian—early Silurian) of the scale was compiled mainly using material from sections of the Siberian platform (RODINOV, 1966; DAVYDOV and KRAVCHINSKY, 1970; KHRAMOV *et al.*, 1965). The Middle-Upper Palaeozoic part was compiled using material from the Russian platform and its margin (BORONIN and BUROV, 1973; MOLOSTOVSKY, 1970; KHRAMOV *et al.*, 1974; SLAUTSITAIS, 1975; KHRAMOV, 1963a, b) and the Altai—Sayan area (APARIN and VLASOV, 1965; POGARSKAYA and KOMISSAROVA, 1977; KHRAMOV *et al.*, 1974) and Tunguska Trough (KHRAMOV *et al.*, 1974).

THE ARGINIAN HYPERZONE (R) occurs at the base of the Palaeozoic section

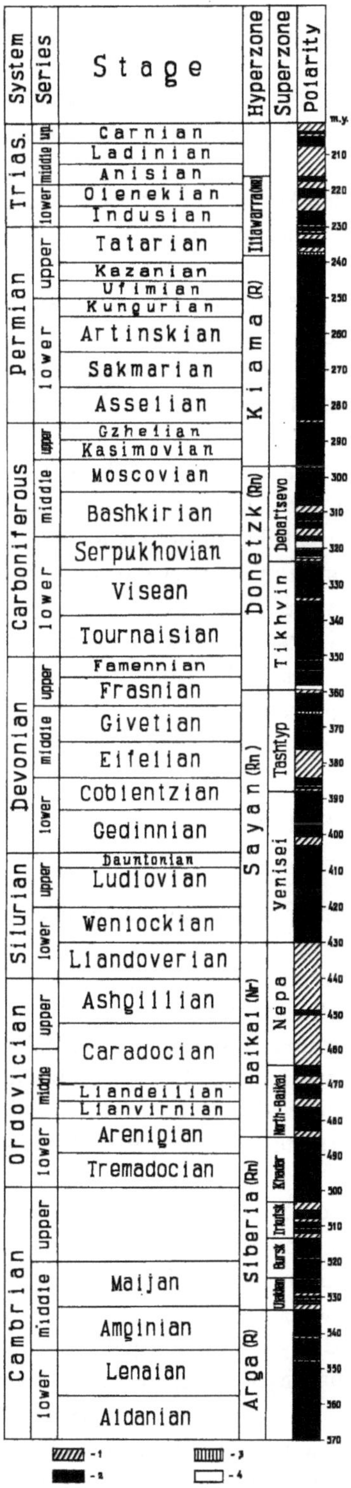

Fig. 1. Polarity scale for the Palaeozoic of the USSR. Magnetostratigraphic units of polarity; 1—normal, 2—reverse, 3—frequent reversals, 4—unknown intervals.

and covers a wide interval from Upper Vendian through the Lower Cambrian to Lower Middle Cambrian.

THE SIBERIAN HYPERZONE (Rn) comprises the remainder of the Middle Cambrian, the Upper Cambrian and most of the Lower Ordovician. It is subdivided into four superzones.

THE BAIKALIAN HYPERZONE (Nr) includes the Middle and Upper Ordovician and apparently most of the Lower Silurian. Two superzones are recognized within this hyperzone.

THE SAYANIAN HYPERZONE (Rn) covers the interval from the uppermost part of the Lower Silurian to the Frasnian stage of Upper Devonian.

THE DONETSK HYPERZONE (Rn) represents the interval from Fammenian to Bashkirian inclusively. The Donetsk hyperzone is subdivided into the Tikhvin and Debal'tsevo superzones.

The boundary between the Sayanian hyperzone and the Donetsk hyperzone is determined mainly by assessing the effect of the metachronous late Palaeozoic remagnetization and the low intensity of the geomagnetic field during this period.

THE KIAMAN HYPERZONE (R) covers the stratigraphic interval from the lower Moscovian (Carboniferous) to the upper Tatarian substage (Permian).

THE ILLAWARRA HYPERZONE (NR). Its lower boundary is placed within the Middle Tatarian (Permian) and the upper boundary lies within the Lower Anisian (Triassic). The Illawarra hyperzone starts with a narrow NR-zone with frequent reversals; 5 R-zones are recognised in the higher parts of many sections.

3. A Comparison of the Palaeozoic Scale of the USSR with World Palaeomagnetic Data

Two main difficulties are encountered when comparing palaeomagnetic profiles on a global scale:

1) commonly there is insufficient stratigraphic coverage so that magnetic zones are incomplete and their boundaries can not be related with certainty.

2) there is no reliable time scale for the Palaeozoic and Mesozoic, especially for stage boundaries. The problems with correlation on a world scale are further compounded because various stratigraphic and geochronological (time) scales are used. Thus a unified palaeomagnetic scale must be regarded as tentative, particularly for intervals of frequent geomagnetic reversals.

Magnetostratigraphic data for regions outside of the USSR for the Early and Middle Palaeozoic are relatively limited e.g. KIRSCHVINK (1978). However, correlation is possible starting from the Debal'tsevo superzone (Fig. 2). Its upper part and boundary with the Kiaman hyperzone was established in North American sections in the Upper Pennsilvanian (MCMAHON and STRANGWAY, 1968) and in the Lower Paganzo of South America (THOMPSON and MITCHELL, 1972). The Kiaman hyperzone, traceable through most continents, is the most complete within these sections. Its upper boundary with the Illawarra hyperzone was reported from the Upper Guadalupian (Permian) of North America (PETERSON and NAIRN, 1971), i.e. at the same level as on the Russian platform.

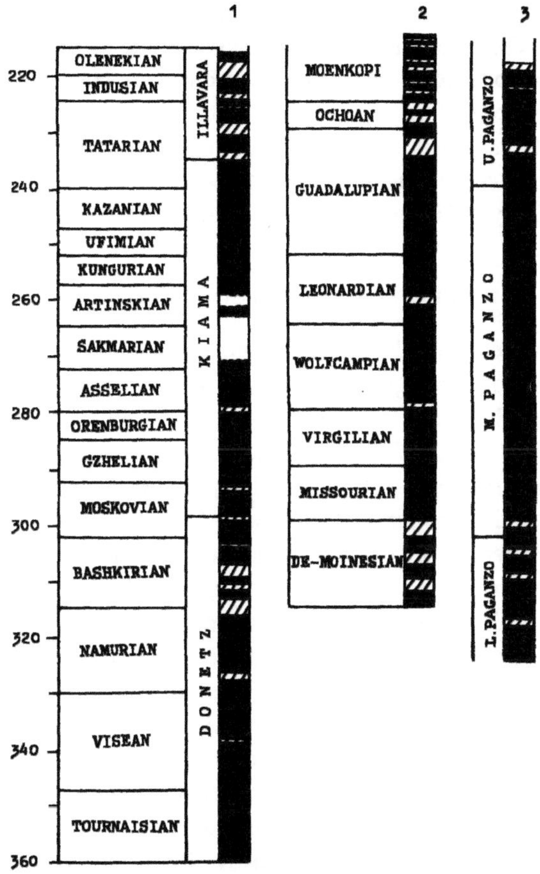

Fig. 2. Correlation of polarity scales for the late Palaeozoic-early Triassic. Scale: 1—USSR, 2—North America (after HELSLEY and STEINER, 1974; MCMAHON and STRANGWAY, 1968; PETERSON and NAIRN, 1971; PICARD, 1964), 3—South America (after THOMPSON and MITCHELL, 1972; VALENCIO *et al.*, 1977).

However, recent results from Permian sections of Central Europe place the boundary between the Kiaman and Illawarra hyperzones at the top of the Saxonian (DACHROTH, 1976): this is considered equivalent to the Kungur stage of the Russian platform. However this correlation is ambiguous. The Illawarra hyperzone has been recorded in the Upper Permian and Lower Triassic sections in many regions of the world. The N- and R-zones can also be correlated with the Russian scale. The Illawarra hyperzone and its upper boundary are best known in sections of the Moenkopi and Chugwater formations (HELSLEY and STEINER, 1974; PICARD, 1964). This boundary is close to the Lower-Middle Triassic boundary (in the USSR it is drawn at the base of Anisian, Fig. 1).

The polarity problem, i.e which of the two poles is the North Pole for the Siberian plate, is evident for the construction of the Early Palaeozoic reversal scale. The scale shown in Fig. 1 is based on the reconstruction characterized by a minimum trajectory of pole migration during the Middle Palaeozoic. However, it is one of two possibilities and is disputed by some workers (ZONENSHAIN, 1976; KROPOTKIN, 1971). Similar problems arise when other plates are discussed and global palaeomagnetic reconstructions for the Early Palaeozoic must be established to resolve these ambiguities. Hence, the problem becomes important for the understanding (a) the early history of geomagnetic reversals and (b) Early Palaeozoic plate tectonics. It can be solved by constructing detailed reversal scales for the continents for the Early and Middle Palaeozoic.

At present the Early Palaeozoic record is known only for the Siberian plate.

4. Cyclic Pattern of the Palaeozoic Geomagnetic Reversals

Hyperzones identified from the palaeomagnetic record Fig. 1, define major stages in the Palaeozoic history of geomagnetic reversals. They are dated as follows: Arginian (R)—590–535 $\times 10^6$ y., Siberian (Rn)—535–485 $\times 10^6$ y., Baikalian (Nr)—485–430 $\times 10^6$ y., Sayanian (Rn)—430–360 $\times 10^6$ y., Donetsk (Rn)—360–300 $\times 10^6$ y., Kiaman (R)—300–240 $\times 10^6$ y., Illawarra (NR)—240–215 $\times 10^6$ y. These units 50–70 $\times 10^6$ y. long are most distinct. There is also shorter rhythmicity of the order 10–40 $\times 10^6$ y. corresponding to superzones. To understand the rhythmicity corresponding to magnetic zones, intervals of one geomagnetic polarity, we can use Fig. 1 to construct a histogram of single polarity time scales (Fig. 3). This histogram has a maximum of about 1–2 $\times 10^6$ y. Thus, the sequence of Palaeozoic magnetic reversals suggests a complex rhythmic pattern yielding times of 5–7 $\times 10^7$, 1–4 $\times 10^7$ and 1–2 $\times 10^6$ y. The analysis of the geomagnetic reversal scale for the entire Phanerozoic shows that rhythms with times of 5–7 $\times 10^7$ y. and longer geomagnetic stages, marked by the commencement of long-term epochs of frequent reversals over about 2 $\times 10^8$ y., may be correlated with patterns of global tectonism, tectonomagmatic cycles, changes in movements of lithospheric plates and changes in marine transgressions and regressions (KHRAMOV, 1970; KHRAMOV *et al.*, 1965; VOGT,

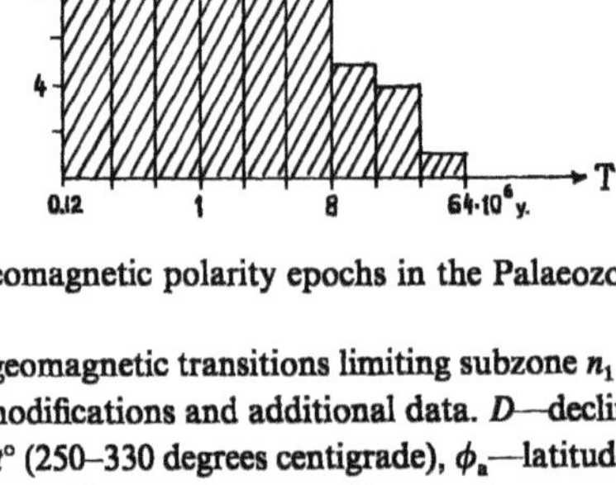

Fig. 3. Duration of geomagnetic polarity epochs in the Palaeozoic and early Triassic.

Fig. 4. Late Permian geomagnetic transitions limiting subzone $n_1 R_3 P$ at Yug River. After BUROV *et al.*, (1977) with modifications and additional data. *D*—declination, *J*—inclination of NRM after cleaning with $t°$ (250–330 degrees centigrade), ϕ_a—latitude of virtual geomagnetic pole in palaeogeographic coordinates, corresponding to a mean pole $\phi=48$, $\Lambda=165$, Q_0—Koenigsberger ratio for NRM after cleaning.

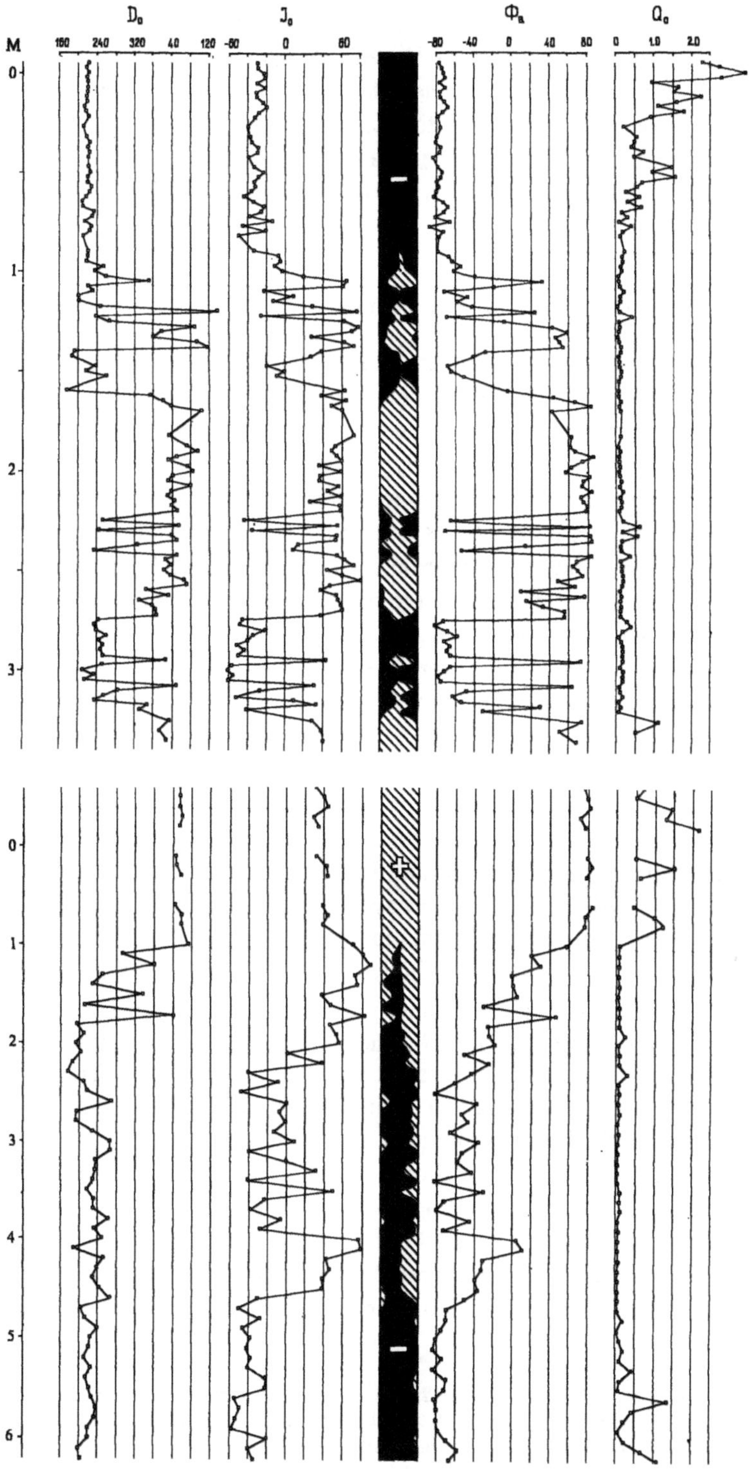
D₀
J₀
Φₐ
Q₀
M
160
240
320
40
120
-80
0
80
-80
-40
0
40
80
0
1.0
2.0
0
1
2
3
0
1
2
3
4
5
6

Fig. 4

1975). This correlation may indicate a relationship between deep-seated tectonic processes and endogenous processes operative in the earth's interior (SHEINMANN, 1975).

The length of time for which a single polarity period existed in the Palaeozoic (Fig. 3) is greater compared with values for the Late Cenozoic. However studies of the fine structure of the geomagnetic field in the palaeozoic (BUROV *et al.*, 1977; KHRAMOV *et al.*, 1974) shows that the difference is mainly due to relatively poor sampling of the Palaeozoic sections.

5. Morphology of Early Palaeozoic Geomagnetic Reversals and Secular Variation

The cause of geomagnetic reversals remains puzzling. A solution could give the clue to a physical theory of the earth's magnetism and help to decipher processes operative within the earth's core. Five different reversals, indexed according to the classification adopted in the USSR as $N_1P \to R_2P$, $R_2P \to N_2P$, $r_1R_3P \to n_1R_3P$, $n_1R_3P \to r_2R_3P$, and $R_1T \to N_2T$ (BUROV *et al.*, 1977, 1979) were studied in the Upper Tatarian and Lower Triassic red bed sections in the eastern Russian platform, at coastal outcrops of the Volga, Vyatka and Yug rivers and in the Transkaspian region on the Mangyshlak Peninsula.

The early Palaeozoic reversals of the geomagnetic field were studied in exposures along the Lena River, in the southern Siberian platform. The results are based on the magnetostratigraphic record from several sections.

Features of the late Permian reversals are exemplified by the two transitions limiting the n_1R_3P—subzone, it occurs in the uppermost Permian zone of reversed magnetization. The two transitions were studied in coastal outcrops where the Yug River joins the Luza River (BUROV *et al.*, 1977). The results are shown in Fig. 4.

Plots of three angular components, declination D, inclination J and palaeolatitude of the virtual geomagnetic pole ϕ_a, show that the transition of the field from one polarity to another polarity is rhythmic and only occasionally does it return to its original polarity. Such rhythms, estimated from sedimentation rates, have a particular time of the order $n \times 10^3$ y. One can also see a shorter rhythmicity of field variations of the order $n \times 10^2$ y. It is clear that a reversal took place against a background of a weaker field (its relative intensity was estimated from $Q_0 = I_n^0/\chi H'$, where I_n^0 is NRM after cleaning, χH is inductive magnetization. The two successive transitions are similar, they display comparable changes in angular parameters and in their patterns of "palaeontensity," Q_0, though representing opposite field transformations.

The details of the reversals are shown in the map of the virtual geomagnetic pole displacement (Fig. 5). It is interesting to note that they differ significantly in the palaeologitudinal distribution of virtual poles.

All the late Permian and early Triassic reversals studied have certain features in common (BUROV *et al.*, 1977, 1979). Transitions begin with a decrease in the geomagnetic

Fig. 5. Virtual geomagnetic poles during the late Permian reversals (relative to palaeogeographic coordinates; longitudes are calculated assuming permanent longitude of the centre of land masses). a–d—successive parts of trajectory of the VGP during the reversal $r_1R_3P \to n_1R_3P$, e—distribution of VGPs for this reversal, *f*—the same for the reversal $n_1R_3P \to r_2R_3P$.

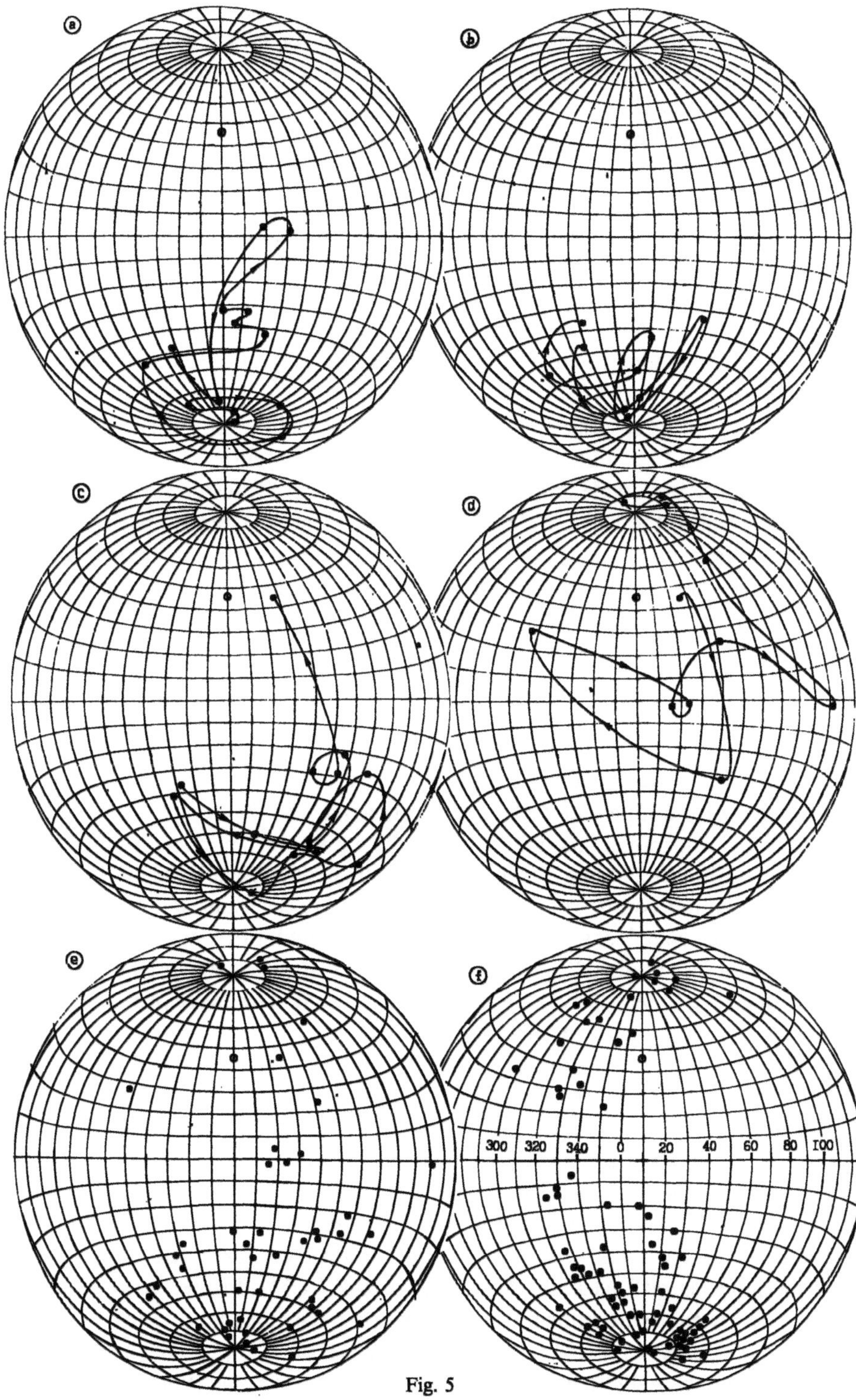

Fig. 5

field intensity (1/4 to 1/10 of the original) and are terminated with a return to original intensity values. Sometimes, preceding the drop in intensity, it exhibits an increase (2 to 4 times); a similar effect is also observed at the beginning of a new polarity stable field interval (BUROV *et al.*, 1977).

When the earth's field decreases, this is accompanied by a loss of direction stability and a "sudden" field reversal takes place. The subsequent rhythms exhibit sharp and complete field reversals. When the former polarity is restored (though sometimes incompletely) it is accompanied by a gradual decrease in amplitude of angular component oscillations; they acquire values typical of the new field polarity. The two reversals shown in Fig. 5 demonstrate a transition in 3–5 rhythms; in other cases the last "smooth" transition stage ($R_2P \rightarrow N_2P$) is absent or only a fraction of the "rhythm" represents the entire reversal; it is a sudden change with a very short duration of decreased field intensity ($N_1P \rightarrow R_2P$ and $R_1T \rightarrow N_2T$) (BUROV *et al.*, 1979).

The early Palaeozoic transitions were investigated mainly using D, J, ϕ_a values; the ratio I_n^0/I_{rv} was used to evaluate relative changes in field intensity.

The transitions are divided into two types viz. a smooth transition from one polarity to another, typical of the late Cambrian; and sharp reversals characteristic of the Early and Middle Ordovician (KHRAMOV *et al.*, 1974). In general a geomagnetic field transition is a complex process involving a transformation from one state to another. The field stabilizes when several reversals have occurred.

Both instant (relatively fast) and smooth reversals were found in the late Cambrian whilst the Ordovician is characterized exclusively by a succession of instant reversals (Fig. 6).

The morphology of the transitions is demonstrated in the trajectories of the virtual geomagnetic poles. The coordinates were calculated relative to the palaeogeographic reconstruction for the Late Cambrian and Ordovician (Fig. 7).

The Ordovician transitions, unlike those of Late Cambrian, show a series of sharp field reversals without passing through intermediate directions. One particular early Ordovician transition consists of a number of field reversals separated by intervals of a short term stable N-polarity. The normal polarity field prior to final stabilization demonstrates an "anomalous" mean direction (relict displacement of the geomagnetic axis) and a high (up to 60 degrees) amplitude of variations. The entire process takes place at one order of magnitude lower field intensity.

Middle Ordovician transitions are of the same type as the Early Ordovician transition and differ only in the number of rhythms.

Although a VGP pattern during a transition may be distorted due to the post-depositional or chemical magnetizations (BAAG and HELSLEY, 1974), this is not considered a problem with results described here since transition pattern details are comparable between sections. However, the question of the persistency of the dipole field during the Palaeozoic reversals remains open. The distribution of VGP's in a restricted path along an ancient meridian as well as multiple returns of the VGP's during the reversal whilst remaining consistent with known Permo-Triassic palaeomagnetic poles (Fig. 5) are regarded as evidence in favour of the persistence of a dipole field. Similar field configurations are characteristic of the early Palaeozoic reversals (Fig. 7).

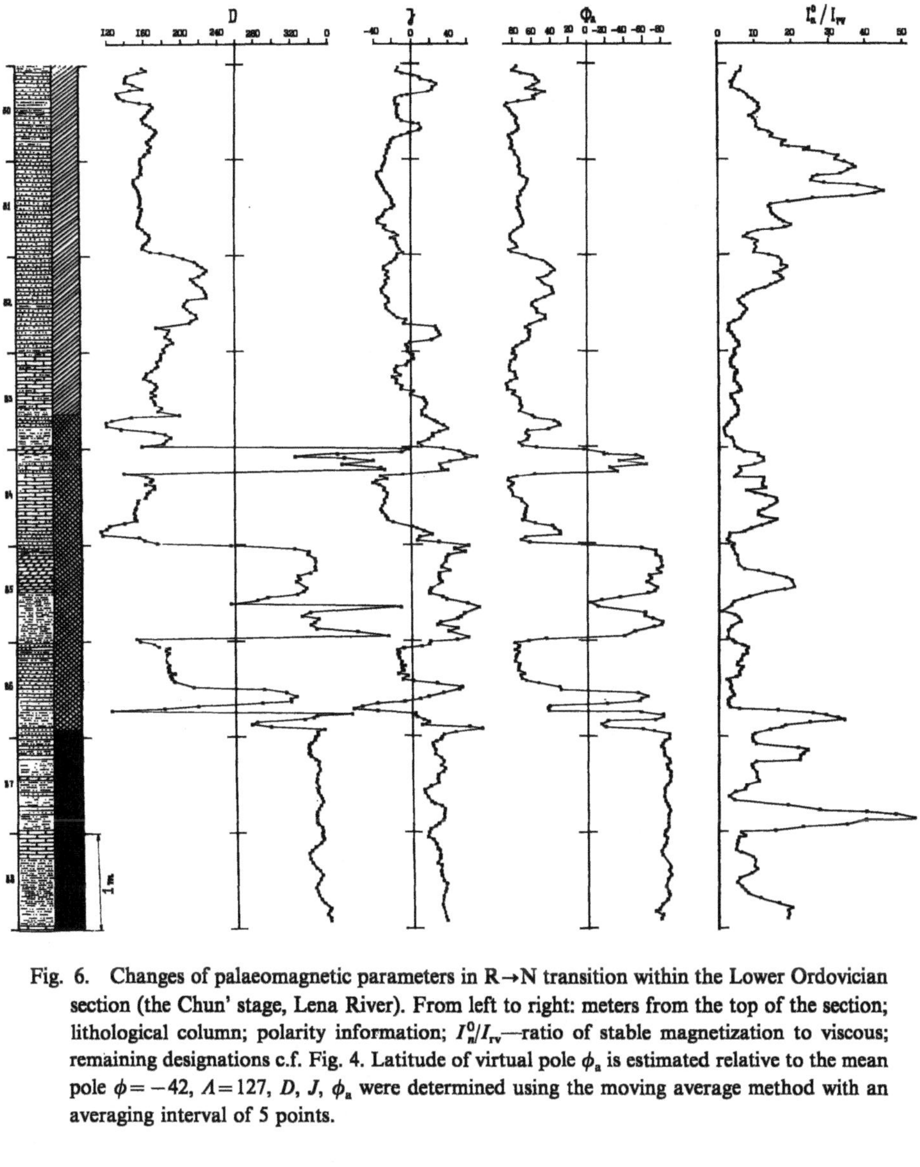

Fig. 6. Changes of palaeomagnetic parameters in R→N transition within the Lower Ordovician section (the Chun' stage, Lena River). From left to right: meters from the top of the section; lithological column; polarity information; I_n^0/I_{rv}—ratio of stable magnetization to viscous; remaining designations c.f. Fig. 4. Latitude of virtual pole ϕ_a is estimated relative to the mean pole $\phi=-42$, $\Lambda=127$, D, J, ϕ_a were determined using the moving average method with an averaging interval of 5 points.

Specific features of some of these reversals were traced over a distance of 500 to 1600 km (KHRAMOV *et al.*, 1974; BUROV *et al.*, 1977). This determines a lower limit for the number of cells into which the field could transform during reversals.

The net time for a geomagnetic reversal to occur during the last 5 m.y. of the Palaeozoic is $1–2\times10^5$ y., i.e. 2–4% of the time. Such an "index of palaeomagnetic disturbance" may be a characteristic feature of certain geological epochs. Similarly, at the close of the Cambrian, it has been estimated to cover about 10 to 15% of the time (KHRAMOV, 1978). This parameter may be significant in a study of the evolution of reversals through geological time.

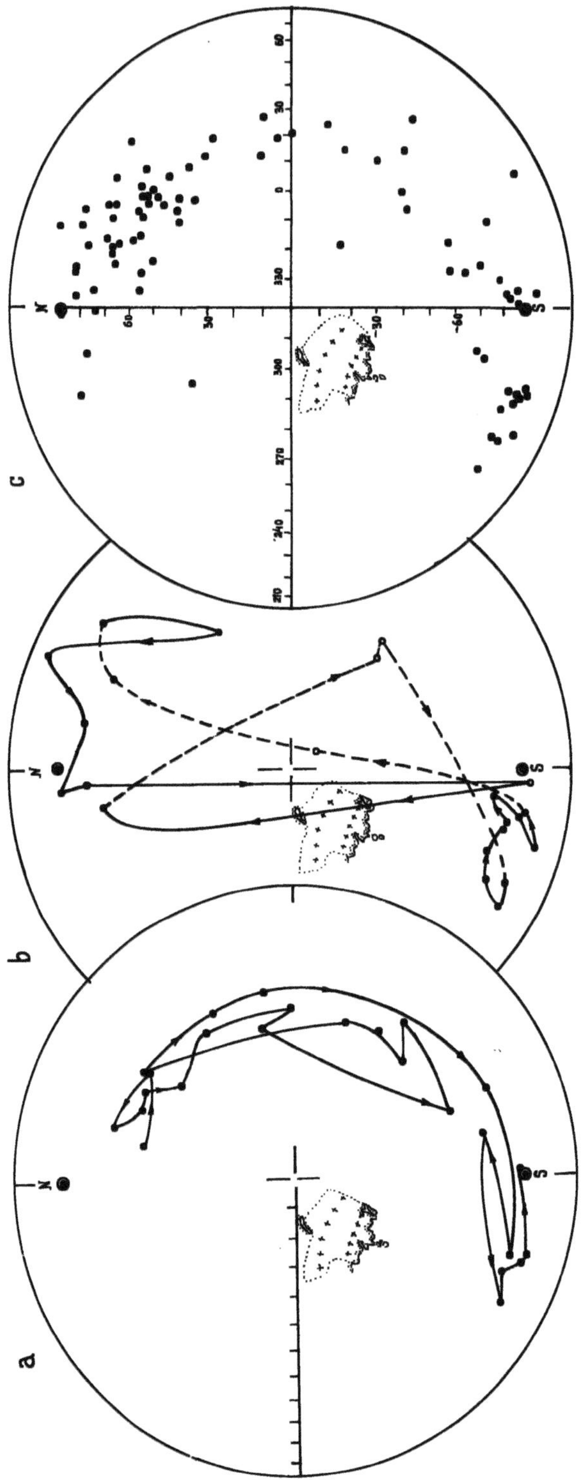

Fig. 7. Trajectory of VGPs during the late Cambrian transitions: Palaeogeographic coordinates, pole $\phi = -36$, $\Lambda = 126$; a—the reversal $n_5 \rightarrow r_1$, at Khamra village; b—excursion (two closed reversals) during the epoch "r_1", at Polovinka village; c—distribution of poles for the $n_5 \rightarrow r_1$ reversal from other sections.

6. Secular Variation of Angular Components of the Geomagnetic Field in the Palaeozoic

Palaeosecular variation studies have revealed that the amplitude characteristics of the variations are more easily obtained than frequency characteristics. Irregular and discontinuous patterns of sedimentation and the asynchroneity of magnetization, common problems in studies of reversals, make it difficult to establish the high-frequency components of secular variation with periods less than 10^4 y. (KHRAMOV, 1967; POGARSKAYA, 1973; KHRAMOV *et al.*, 1974).

Studies of amplitude characteristics of palaeosecular variation during the Palaeozoic were carried out on sections of Carboniferous and Permian sedimentary rocks of the Donetsk basin (KHRAMOV, 1967; POGARSKAYA, 1973). Amplitudes were estimated from the standard angular deviation of the field direction, δ, and were calculated from vector dispersion in a two-tier analysis (IRVING, 1964). It has been established (POGARSKAYA, 1973), that experimental data fall on the dispersion versus latitude curve estimated for the effects of the non-dipole components of the recent geomagnetic field (IRVING, 1964). This implies that both the relative contribution of non-dipole components and the degree of secular variation in the late Palaeozoic were similar to those operative during recent times.

A frequency spectrum of secular variation in the Palaeozoic was determined from a comparison of amplitudes using contemporaneous deposits, and accounting for their different rates of sedimentation. Palaeosecular variation cycles with $T \geqq 10^4$ y. and angular standard deviations, $\delta = 7$–10 degrees (KHRAMOV, 1967), were detected.

Long period secular variation cycles within the Kiamian interval were found in two parallel sections of the Upper Kazanian red beds in Tartaria. The period of variations is of the order 10^5 y. (BORONIN, 1979).

Four our studies we have used a continuous sampling technique within sections and have separated short- and long-term palaeosecular variation components by calculating a running mean. The analysis suggests the presence of a discrete variation spectrum. The shortest period component is regarded as noise and since its record from sections is fragmentary it cannot be isolated. This component was therefore excluded from plots of secular variation (D and J) by calculating the moving average. The results from such treatment are shown in Fig. 6.

Secular variation of the geomagnetic field during the unstable polarity epoch at the close of the Cambrian is represented by complex VGP loops. The predominant rotation direction of the virtual geomagnetic axis is counterclockwise when looking from the north and is independent of geomagnetic polarity (Fig. 8).

Long-period variations of the geomagnetic field during stable periods of both polarities in the late Cambrian and early Ordovician are revealed after smoothing of angular values D, J, ϕ_a and the scalar quantity I_n/I_{rv}. The amplitude of angular variations during periods of reversed polarities reaches 10 degrees with a period of 4×10^4 y.

Short-period directional variations of the field within an averaging intervals of 10^4 y. yield standard deviations of 6–15 degrees. Variations for 3–4×10^4 y. are traceable using the ratio I_n^0/I_{rv}, and it varies sinusoidally along the section though the amplitude is not constant.

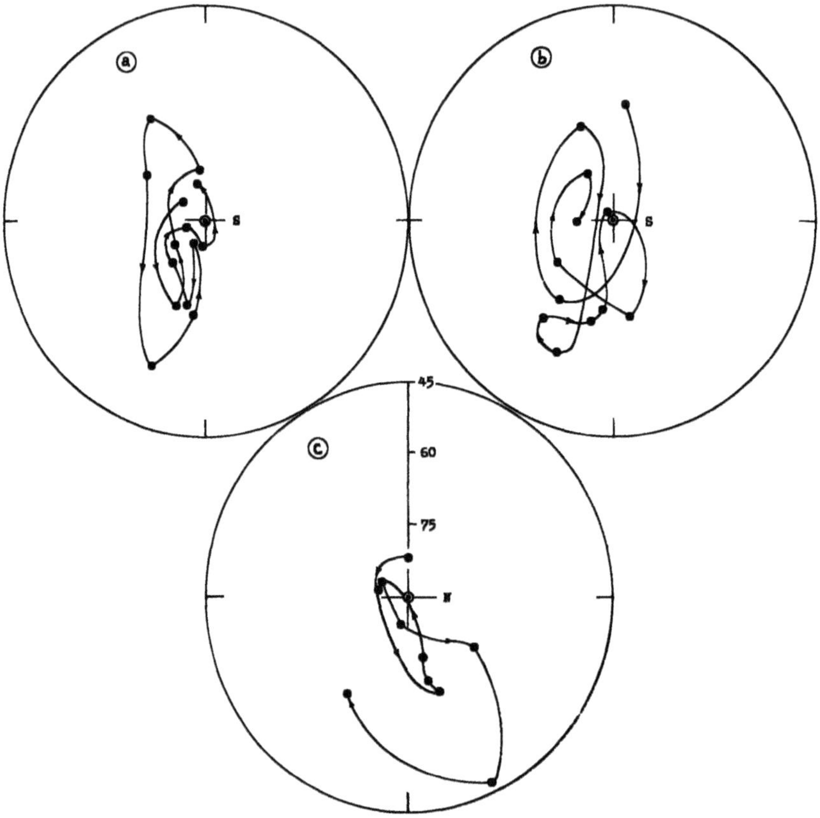

Fig. 8. Trajectory of VGPs corresponding to secular variations in the early Palaeozoic (palaeogeographic coordinates). Late Cambrian: a—before the $n_5 \rightarrow r_1$ reversal, Zmeinovo village; b—just after the $n_5 \rightarrow r_1$ reversal, Polovinka village; c—at the beginning of the epoch r_1, Khamra village.

Figures 4 and 6 show that during periods of frequent geomagnetic reversals, e.g. the end of the Cambrian, the beginning of the Ordovician and the end of Permian, secular variation was complex and comprised components with periods of $T \leq 10^4$ y., $T_2 = n \times 10^4$ y. and $T_3 = m \times 10^5$ y. estimated from sedimentation rates. The values of n and m vary between 1 and 3. Angular standard deviations are $\delta_1 = 6$–15 degrees, $\delta_2 = 7$–10 degrees and $\delta_3 \approx 10$ degrees corresponding to amplitudes $A_1 = 18$–45 degrees, $A_2 = 20$–30 degrees, $A_3 = 30$ degrees, respectively.

Our interpretation is that the first harmonics are produced by changes (drift?) of the non-dipole field, the latter reflects variations of the main dipole. We have used a model of secular variation M which describes the field for the last 5 m.y. (McElhinny and Merrill, 1975). Within the accuracy of the measurements the parameters for the model have not changed greatly since the early Palaeozoic.

7. Model of the Central Axial Dipole for the Palaeozoic

To test the applicability of the central axial dipole model to the Palaeozoic requires the consideration of two main points: firstly, to what extent are the angular elements of the palaeomagnetic field over the earth's surface consistent with a central axial dipole? Secondly, what is the origin of peculiarities of the ancient geomagnetic field at specific observation points and do they represent real deviations from the dipole pattern?

The structure of a palaeomagnetic field for ancient epochs depends primarily on the palaeogeographic base that was used. Such a base reliably derived independent of palaeomagnetic data is available only for the end of Palaeozoic, e.g. Wegener's Pangaea. Using this base-map spherical harmonic analyses showed that the palaeomagnetic field for the Triassic, Permian and Carboniferous approximate well (with an error of 7–10 degrees) to the dipole field, oriented along the earth's rotation axis but displaced from the geocenter by 820–870 km (Benkova *et al.*, 1973). On this model the magnetic centre of the earth is always displaced towards the Pacific Ocean. This assymmetry (if real) is crucially important for understanding the earth's magnetism.

Spherical harmonic analysis of the main field and our knowledge of the characteristics of palaeosecular variation during the Palaeozoic and recent epochs constrain us to accept the central axial dipole model as a working basis for the interpretation of palaeomagnetic data for the Palaeozoic.

The features described from many Palaeozoic reversals, e.g. the long-term displacement of the mean virtual geomagnetic axis near reversals and the considerable duration of the transition, lead us to conclude that periods of systematic displacement of palaeomagnetic directions may occupy rather long intervals within sedimentary sections. An entire volcanic series may fall within this time-span. Long-period secular variations produce a similar effect. Thus, the problem of time averaging acquires significance at least for some Palaeozoic epochs. The application of the central axial dipole model is apparently possible only by securing an averaging interval up to values of the order of 1 m.y.

A comparison of normal and reverse magnetization directions from the early Palaeozoic of Siberia suggests the presence of an irreversible equatorial dipole during periods of frequent reversals. The projection of this dipole lies on the great circle close to the trajectory of the virtual geomagnetic pole studied in detail for the Late Cambrian reversals (Fig. 7). Such a coincidence suggests that the irreversible dipole influences the morphology of the transitions characterized by a smooth loop-like change of angular components.

The authors are very grateful to V. B. Burov (Kazan State University), M. T. Nemova and S. A. Pisarevsky (VNIGRI) who took part in field and laboratory studies; G. N. Petrova, D. M. Pechersky (Institute for the Physics of the Earth) and E. A. Molostovsky (Saratov University) for useful discussions at different stages of the work, I. P. Slautsitais (VNIGRI) for help with the preparation of the manuscript. The authors are especially grateful to R. E. Sorkina (Institute of the Geology of the Arctic) for the English translation of the manuscript.

REFERENCES

Aparin, V. P. and A. Ya. Vlasov, Nekotorye voprosy istorii geomagnitnogo polya v konze srednego paleozoya i v verkhnem paleozoe, in *Nastoyashchee i proshloye magnitnogo polya Zemli,* pp. 213–220, Nauka, Moscow, 1965.

BAAG, C. and Ch. E. HELSLEY, Evidence for penecontemporaneous magnetization of the Moenkopi formation, *J. Geophys. Res.*, **79**, 3308–3320, 1974.

BENKOVA, N. P., A. N. KHRAMOV, T. N. CHEREVKO, and N. V. ADAM, Spherical harmonic analyses of the palaeomagnetic field, *Earth Planet. Sci. Lett.*, **18**, 141–147, 1973.

BORONIN, V. P., Thin structure of geomagnetic field: secular variations and reversal processes, in *The Earth Core on Geomagnetic Data,* Part I, edited by V. Bucha and G. N. Petrova, 204 pp., Praha, 1980.

BORONIN, V. P. and B. V. BUROV, O sootnoshenii stratigrapficheskikh i paleomagnetnykh granits v razrezakh tatarskogo yarusa permi i nizhnego triasa Srednego Povolzhia, in *Postoyannoe magnitnoe pole Zemli, paleomagnetism i magnetism gornykh porod,* pp. 80–82, Naukova Dumka, Kiev, 1973.

BUCHA, V., The continuous pattern of variations of the geomagnetic field in the Quaternary and their causes, *Studia Geophys. Geodaet.*, **117**, 218–231, 1973.

BUROV, B. V., R. A. KOMISSAROVA, I. P. SLAUTSITAIS, and A. N. KHRAMOV, Geomagnitnye inversii na rubezhe paleozoya i mesozoya, in *Paleomagnetism i voprosy tektoniki plit,* pp. 13–26, Trudy VNIGRI, Leningrad, vyp. 394, 1977.

BUROV, B. V., I. P. SLAUTSITAIS, and A. N. KHRAMOV, Thin structure of geomagnetic field: secular variations and reversal processes, in *The Earth Core on Geomagnetic Data,* Part I., edited by V. Bucha and G. N. Petrova, 204 pp., Praha, 1980.

COX, A., Geomagnetic reversals, *Science*, **163**, 237–245, 1969.

CREER, K. M., D. L. GROSS, and J. LINEBACK, Origin of regional magnetic variations recorded by Wisconsin and Holocene sediments from lake Michigan, USA, and lake Windermere, England, *Geol. Soc. Am. Bull.*, **87**, 531–540, 1976.

DACHROTH, W., Gesteinsmagnetische Marken in Perm Mitteleurope. Geolog. Jahrb., E, N 10, pp. 1–71, Hannover, 1976.

DAVYDOV, V. F. and A. Ya. KRAVCHINSKY, Paleomagnitnye issledovania stratifizirovannykh porod Vostochnoi Sibiri, in *Geofizicheskie issledovania pri reshenii geologicheskikh zadach v Vostochnoi Sibiri.*, pp. 124–147, Nedra, Moscow, vyp. 4, 1970.

HELSLEY, Ch. E. and M.B. STEINER, Paleomagnetism of the lower Triassic Moenkopi formation, *Geol. Soc. Am. Bull.*, **85**, 457–465, 1974.

IRVING, E., *Palaeomagnetism and its application to geological and geophysical problems*, 399 pp., John Wiley, New York, 1964.

KHRAMOV, A. N., Paleomagnethye issledovania verkhnego paleozoya i triasa zapadnoi chasti Donbassa, in *Paleomagnithye stratigraficheskie issledovania,* pp. 97–117, Gostoptekhizdat, Leningrad, 1963a.

KHRAMOV, A. N., Paleomagnitnoe izuchenie razrezov verkhnei permi i nizhnego triasa severa i vostoka Russkoi platformy, in *Paleomagnitnye stratigraficheskie issledovania*, pp. 145–174, Gostoptekhizdat, Leningrad, 1963b.

KHRAMOV, A. N., Magnitnoe pole Zemli v pozdnem paleozoe, *Izv. Akad. Nauk SSSR, ser. Fizika Zemli,* N 1, pp. 86–108, 1967.

KHRAMOV, A. N., Horizontal movements in the Earth's crust and palaeomagnetism, *Phys. Earth Planet. Inter.*, **2**, 367-372, 1970.

KHRAMOV, A. N., Paleomagnitnaya shkala verkhnego paleozoya Donbassa i Priuralia i ee znachenie dlya problemy mezhkontinentalnykh korrelatsy, in *Stratigraphia karbona i geologia uglenosnykh formatsy SSSR,* pp. 31–38, Nedra, Moscow, 1975.

KHRAMOV, A. N., Paleomagnetism i problemy geotektoniki, in *Tektomosfera Zemli,* pp. 280–302, Nauka, Moscow, 1978.

KHRAMOV, A. N., V. P. RODIONOV, and R. A. KOMISSAROVA, Novye dannye o paleozoiskoi istorii zemnogo magnitnogo polya na territorii SSSR, in *Nastoyashchee i proshloe zemnogo magnitnogo polya*, pp. 206–213, Nauka, Moscow, 1965.

KHRAMOV, A. N. (Ed)., G. I. GONCHAROV, R. A. KOMISSAROVA, I. A. POGARSKAYA, V. P. RODIONOV, I. P. SLAUTSITAIS, L. S. SMIRNOV, and N. N. FORSH., *Paleomagnetism paleozoya*, 236 pp., Nedra, Leningrad, 1974.

KIRSCHVINK, J. L., The Precambrian-Cambrian boundary problem: magnetostratigraphy of the Amadeus basin, central Australia, *Geolog. Magaz.*, **115**, 139–150, 1978.

KROPOTKIN, P. N., Eurasia as a composite continent, *Tectonophysics*, **12**, 261–266, 1971.

LARSON, R. L. and C. PITMAN, World-wide correlation of Mesozoic magnetic anomalies and implications, *Geol. Soc. Am. Bull.*, **83**, 3645–3662, 1972.

MCELHINNY, M. W. and R. T. MERRIL, Geomagnetic secular variation over the past 5 m.y., *Rev. Geophys. Space Phys.*, **13**, 687–708, 1975.

MCMAHON, B. E. and D. W. STRANGWAY, Investigation of Kiaman magnetic division in Colorado red beds, *Geophys. J. R. A. S.*, **15**, 265–285, 1968.

MOLOSTOVSKY, E. A., Paleomagnithye dannye po stratigrapfii verkhnei permi i triasa vostochnykh raionov Russkoi platformy i Preduralskogo kraevogo progiba, in *Mat. VII konf. po postoyannomu geomagn. polyu i paleomagnetismu*, II, pp. 98–101, Naukova Dumka, Kiev, 1970.

MOLOSTOVSKY, E. A., M. A. PEVZNER, D. M. PECHERSKY, V. P. RODIONOV, and A. N. KHRAMOV, Magnitostratigraficheskaya shkala fanerozoya i rezhim inversii geomagnitnogo polya, in *Geomagnitnye issledovania*, N 17, Nauka, Moscow, 1976.

NAKAJIMA, T. and N. KAWAI, Secular geomagnetic variation in the recent 60,000 years found from the Lake Biwa sediments, in *Rock magnetism and paleogeophysics*, Vol. 1, pp. 34–38, Publ. by the Rock Magnetism and Paleogeophysics Res. Group in Japan, Tokyo, 1973.

PETERSON, D. N. and A. E. M. NAIRN, Palaeomagnetism of Permian red beds from the SW USA, *Geophys. J. R.A.S.*, **23**, 191–207, 1971.

PICARD, M. D., Paleomagnetic correlation of units within Chugwater (Triassic) formation, *AAPG Bull.*, **48**, 269–291, 1964.

POGARSKAYA, I. A., Dannye o vekovykh variatsyakh na rubezhe karbona i permi, in *Postoyannoe magnitnoe pole Zemli, paleomagnetism i magnetism gornykh porod*, pp. 43–45, Naukova Dumka, Kiev, 1973.

POGARSKAYA, I. A. and R. A. KOMISSAROVA, Paleomagnetism srednego paleozoja Russkoi i Sibriskoi platform, in Paleomagnetism i voprosy tektoniki plit, pp. 63–107, Trudy VNIGRI, vyp. 394, 1977.

POSPELOVA, G. A., Ob osobennostyakh povedenia geomagnitnogo polya v pliozen-chetvertichnoe vrem y a, *Doklady Akad. Nauk SSSR*, **210**, 663–665, 1973.

RODIONOV, V. P., Raschlenenie i sopostavlenie razrezov verkholenskoi svity Irkutskogo amfiteatra po paleomagnitnym dannym, *Geologiya i geofizika*, **6**, 115–128, 1966.

SHEINMANN, Yu. M., Neketorye induktivnye zakonomernosti, vytekayushchie iz paleomagnitnykh izmereni, *Izv. Akad. Nauk, SSSR, ser. Fizika Zemli*, N 1, pp. 66–75, 1975.

SLAUTSITAIS, I.-P., Stratigrapfia permskikh i triasovykh otlozhenyi Mangyshlaka po paleomagnitnym dannym, in *Kontinentalnye krasmozvetnye otlozhenia permi i triasa*, pp. 113–115, Saratov, 1975.

THOMPSON, R. and J. G. MITSHELL, Palaeomagnetic evidence for the age of the lower boundary of the Kiaman magnetic interval in South America, *Geophys. J. R.A.S.*, **27**, 207–214, 1972.

VALENCIO, D. A., J. P. VILAS, and J. E. MENDIA, Palaeomagnetism of a sequence of red beds of the middle and upper sections of Paganzo Group (Argentina) and the correlation of upper Palaeozoic-Lower Mesozoic rocks, *Geophys. J. R.A.S.*, **51**, 59–74, 1977.

VOGT, P. R., Changes in geomagnetic reversal frequency at times of tectonic change: evidence for coupling between core and upper mantle processes, *Earth Planet. Sci. Lett.*, **25**, 355–360, 1975.

ZONENSHAIN, L. P., Rekonstruktsii paleozoiskikh okeanov, in *Dreif kontinentov*, edited by G. N. Petrova, pp. 28–71, Nauka, Moscow, 1976.

Apparent Polar Wander Path of Central Iran and its Geotectonic Interpretation

H. C. SOFFEL* and H. G. FÖRSTER**

*Institut für Allgemeine und Angewandte Geophysik, Universität München, Theresienstrasse, München, West Germany
**Institut für Mineralogie und Lagerstättenlehre, RWTH Aachen, Wüllnerstrasse, Aachen, West Germany

(Received January 28, 1980; Revised July 21, 1980)

Palaeomagnetic investigations have been made on mainly igneous rocks and red sandstones from 167 sites in Central Iran covering the time interval between Quaternary and Late Precambrian. Based on these data an apparent polar wander path has been established for Central Iran which shows marked differences with the apparent polar wander path of Eurasia (IRVING, 1977) but some resemblances with the apparent polar wander paths of the Gondwana landmasses India and Africa (KLOOTWIJK, 1976; MCELHINNY, 1973; IRVING, 1977). The palaeoremanence directions indicate a counter-clockwise rotation of more than 90° of Central Iran with respect to Eurasia since Upper Paleozoic. According to the palaeolatitudes and other, mainly geological evidence, Central Iran was part of the Gondwana landmasses throughout parts of the Paleozoic. It was separated much earlier from Gondwana than India and had also a much earlier (possibly Triassic) collision with Eurasia.

1. Introduction

An important question concerning the geotectonic position of Central Iran is its relation to Gondwana and Eurasia. Advocates of fixist concepts claim that Central Iran has always been an integral part of Eurasia (BRUNN, 1976) since Late Proterozoic time. KASHFI (1976) considers even Africa, Arabia, Iran, and India as part of Eurasia, despite uniequivocal palaeomagnetic evidence to the contrary. Other geologists propose that Central Iran has been a microplate which drifted between Gondwana and Eurasia (TAKIN, 1972; DEWEY *et al.*, 1973; STÖCKLIN, 1974). Differences in opinion exist concerning the origin of Central Iran (part of Eurasia according CRAWFORD (1972) or part of Gondwana as suggested by our own early palaeomagnetic studies (BECKER *et al.*, 1973; SOFFEL *et al.*, 1975) and the total amount of drift (splitting away from Arabia or successive northward drift of Gondwana fragments, FÖRSTER, 1976). The palaeomagnetic reconstructions of Gondwana and Eurasia since the early Paleozoic (BRIDEN *et al.*, 1974; IRVING, 1977) give no detailed information about the southern part of Central Asia. This is due to little or no palaeomagnetic information available for Turan, Hindukush, Pamir, Tienshan, Karakorum, and Kuenlun.

With the construction of an apparent polar wander path for Central Iran we hoped to be able to solve the problem of the origin of Central Iran and to provide more detailed information on the drift history of the microplate.

2. General Geological Situation

Our paper follows the structural concept of STÖCKLIN (1977), which is shown in Fig. 1. The Southern Domain (Arabia, India, Zagros, Salt Range, all believed to belong to Gondwana) has a Precambrian basement. Folding occurred only during Pliocene time. The Axial Ophiolite Belt (Zagros Crush Zone, Oman, Makran, all forming the so-called Neotethys) is built of oceanic crust. The Central Domain (Central Iran, Alborz, Lut, Afghanistan, Pamir, all described as Northern Gondwana) has a continental basement. It

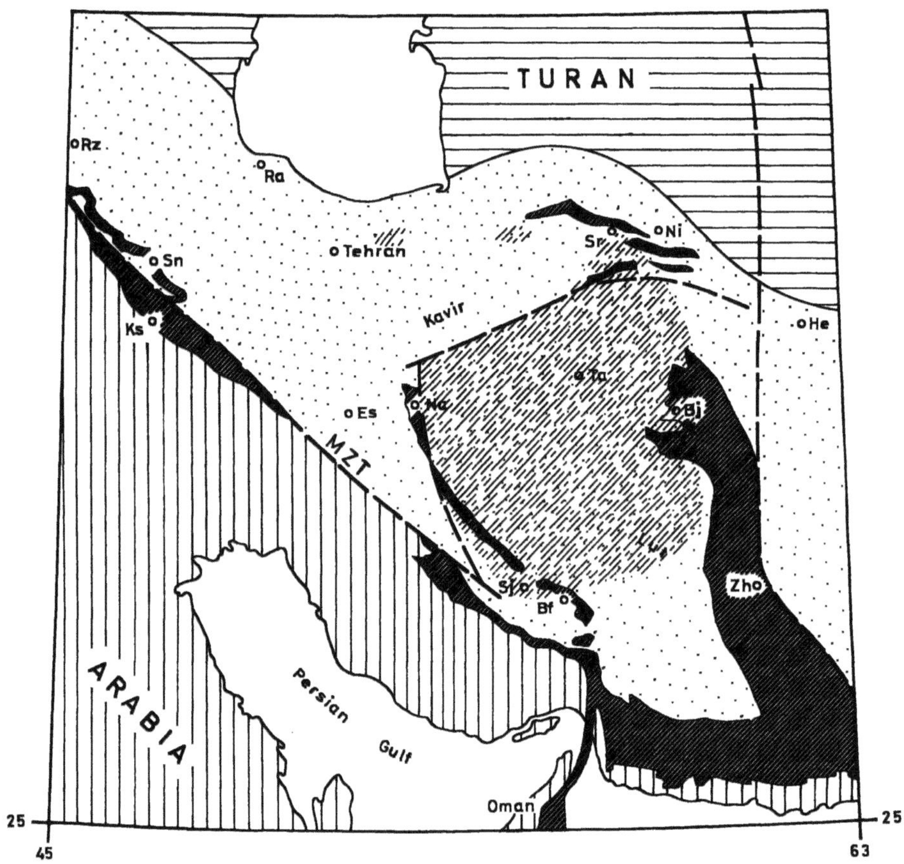

Fig. 1. General structural subdivision for Iran and its surroundings. Southern Domain = vertical lines; Axial Ophiolite Belt = diagonal lines; Central Domain = dotted; Northern Domain = horizontal lines. Ks = Kermanshah, Sn = Sanandaj, Rz = Rezayeh, Ra = Rasht, Es = Esfahan, Na = Nain, Sj = Sirjan, Bf = Baft, Ta = Tabas, Bj = Birjand, Zh = Zahedan, Sr = Sabzevar, Ni = Nishapoor, He = Herat. MZT = Main Zagros Thrust. Sampling is confined to the diagonally hatched areas and is almost exclusively restricted to the Central Iran and Lut Block.

maintained its platform character until Upper Triassic time. The Northern Domain (South Caspian, Kopet Dagh, Turan, Tienshan, all forming the so-called Central Asia and Paleotethys) is built of continental crust consolidated in Hercynian time. It is interesting to note that Central Iran is not correlated with the Himalayas but with the Pamir (STÖCKLIN, 1977).

The history of cratonization of Central Iran is not yet clear. The metamorphic rocks

Table 1. Principle stratigraphy of Central Iran according to STÖCKLIN (1971).

Age	Name of Formation	Lithology		Volcanic Rocks
Pleistocene				
Pliocene	Neogene-Quarternary Volcanics	Conglomerates	VVV	Nephelinbasanite,* Andesite* Latite, Rhyolite*
Oligo/Miocene	Upper Red Formation	Continental Conglomerates*		
	Qom Formation	Limestone		
	Lower Red Formation	Continental Conglomerates	V	Dacite*
Eocene	Eocene Volcanics	Nummulitic Limestone, Flysch	VVV	Basalt,* Phonolite, Trachyte* Andesite,* Rhyolite,* Dacite*
Paleocene/ Upper Cretaceous	Coloured Melange F.	Ophiolites,* Radiolarite* Limestone	VVV	Pillow Lava*
Middle Cretaceous				
Lower Cretaceous/ Upper Jurassic	Garedu F., Shirkuh F.	Red Sandstone*	V	Trachybasalt*
Middle Jurassic	Hojedk Formation	Coal bearing Sandstone*	V	Flood Basalt*
Lower Jurassic	Shemshak Formation	Coal bearing Sandstone*	V	Flood Basalt*
Triassic	Naiband Formation	Limestone*		
	Nakhlak Formation	Sandstone*		
Permian	Abadeh Formation	Shale*		
	Jamal Formation	Limestone*	V	Andesite*
Carboniferous	Sardar Formation	Sandstone*		
Devonian	Bahram F., Shishtu F.	Limestone*		
	Kerman Old Red F.	Red Sandstone*		
Silurian	Niur Formation	Limestone*	V	Tholeiite*
Ordovician/ Middle Cambrian	Mila Formation	Limestone,* Shale*		
Lower Cambrian	Lalun Formation	Red Sandstone*		
Infracambrian	Zaigun Formation	Red Sandstone*		
	Barut Formation	Red Sandstone*		
	Soltanieh Formation	Dolomite*		
	Esfordi, Rizu and Desu F.	Dolomite,* Gypsum	VVV	Rhyolite,* Magnetite*
	Bayandor Formation	Red Sandstone*		
Precambrian	Tashk Formation	Meta-Graywacke*	V	Tuff*
	Crystalline Basement	Gneiss, Amphibolite*		

V: frequent, VVV: very frequent volcanic activities. Within the diagonally hatched area of Fig. 1, favourably igneous rocks and red sandstones have been sampled. The sampled rock types are marked by an asterisk (*).

are similar to those of the shield regions of Northeast Africa and Arabia and India. Therefore the consolidation phase should not be labelled Assyntic or Baikalian but Panafrican (FÖRSTER, 1976). STÖCKLIN (1968) drew attention to the similarity of the Infracambrian-Paleozoic platform cover in Central Iran and Arabia. FÖRSTER (1968) correlated Infracambrian rhyolites from Northeast Iran with the Malani rhyolites in Northwest India (745 ± 10 m.y., CRAWFORD and COMPSTON, 1970). Both rhyolites are conformably overlain by red sandstones grading into fossil bearing Cambrian limestone. A new proof of the Gondwana connection is the discovery of Persimedusites chahgazensis n. sp. in Central Iran near Bafq which according to HAHN and PFLUG (1980) matches metazoons in Southern Austalia (Ediacara Fauna). It should be noted also that the Infracambrian rocks in Central Iran are similar to the Uppermost Sinian in Tarim and Tsaidam (ANONYMOUS, 1976). Both rock series contain Medusinites, Hyolithoidae, Collenia, and microplants. Therefore we would like to include the Tarim and Tsaidam into the Central Domain as well, although STÖCKLIN (1977) places them into the Northern Domain.

The principal stratigraphy of Central Iran is given in Table 1 (according to STÖCKLIN, 1971). It also shows the presence (V) of horizons with volcanic rocks and formations with abundant volcanic activities (VVV). A detailed map of the structural geology of the Iran has been published by STÖCKLIN (1977). Our sampling is confined to the diagonally hatched area in Fig. 1.

3. Palaeomagnetic Sampling and Measurements

Sampling was made between 1972 and 1977 in the region shown in Fig. 1 using portable drilling equipment and taking at least 7 cores at each of the 167 sites. The

Table 2. Mean pole positions (λ°E, φ°N) for the different formations.

Age	Number of sites	N	R	k	α_{95}(°)	λ (°E)	φ (°N)
T_u	8	4	3.94	50.81	9.82	174.1	76.8
T_m	7	5	4.71	13.96	16.75	302.6	64.1
T_1	36	28	26.13	14.47	6.96	310.6	62.0
K	23	12	11.48	21.19	8.78	287.6	78.5
J	20	11	10.18	12.17	12.10	314.4	38.0
T_r	3	3	2.88	17.15	19.52	348.8	15.2
P	5	4	3.53	6.36	27.75	306.8	34.7
C	3	0					
D	6	5	4.59	9.78	20.02	276.6	66.0
S	6	0					
O	5	0					
Є	7	0					
IЄ	38	18	14.87	5.44	14.15	209.0	35.5

The total number of sites sampled for each formation is given as well as the number of sites used for the computation of the mean together with other statistical parameters: residual vector R; precision parameter k; α_{95}, semiangle of 95% cone of confidence.

numbers of sites taken for the different formations are listed in Table 2. Only the Infracambrian and post Triassic rocks are sufficiently well sampled. Sampling of Paleozoic and Lower Mesozoic rocks is still unsatisfactory. The reason for this is given in Section 6. The different rock types sampled are marked in Table 1 by an asterisk.

The limited space available in this paper does not permit us to present for all 167 sites a description of the locality name, the geographical coordinates, the local geological situation, NRM direction and statistical parameters, bedding correction, demagnetization behaviour and special treatment, characteristic remanent magnetization and its statistical parameters, pole positions. However, these data are available on Tables and can be obtained from the authors upon special request.

3.1 Natural remanent magnetization (NRM)

For all sites the NRM was measured with a shielded Digico spinner magnetometer. The standard sized specimens (1 inch in diameter, 2.4 cm long) were stored randomly in part for several weeks in the laboratory before the measurements were started. So the amount of scatter of NRM especially of sister specimens from one core was already indicative for the capability of a certain rock type to acquire viscous components of remanence (VRM) during the present Brunhes epoch.

Only 3 sites showed NRM intensities (which did not increase during demagnetization) less than 10^{-8} Gauss and were therefore not measurable with our instrument (a limestone, a sandstone and a granite). Two iron ores of Infracambrian age showed extremely high magnetic viscosity. In many cases the NRM had little scatter and the directions grouped away from the local present field direction. However for a large number of sites the NRM grouped with different degrees of scatter around the local present field direction or was so scattered that only mean directions with low values for the precision parameter ($k<3$) could be determined.

3.2 Demagnetization studies

At least two pilot specimens from each site, regardless of the rock type, were subjected to AF demagnetization up to 2,000 Oe. The changes in directions and intensities were analysed using Zijderveld diagrams (see Fig. 2) and stereoplots of the directions alone for the detection of great circle distributions. Evidence for multicomponent remanence was encountered as expected especially for the Pre-Tertiary rocks. In general the softer component was aligned parallel to the local geomagnetic field at the sampling site or was more or less randomly scattered around this direction. For the final AF demagnetization the proper field intensity for the isolation of the most stable remanence component was used. According to the general experiences in palaeomagnetism, this direction can in general be related to the age of the rocks, especially when a positive fold test can be demonstrated (see Section *3.3*) and no other younger geological event can be responsible for a total magnetic overprint. However, in some cases the fold tests were negative and no other consistent remanence component could be isolated. These sites were then omitted from any further statistical treatment.

After AF demagnetization and the removal of the softest remanence components at least two specimens of all sites of the sedimentary and of some critical sites of igneous

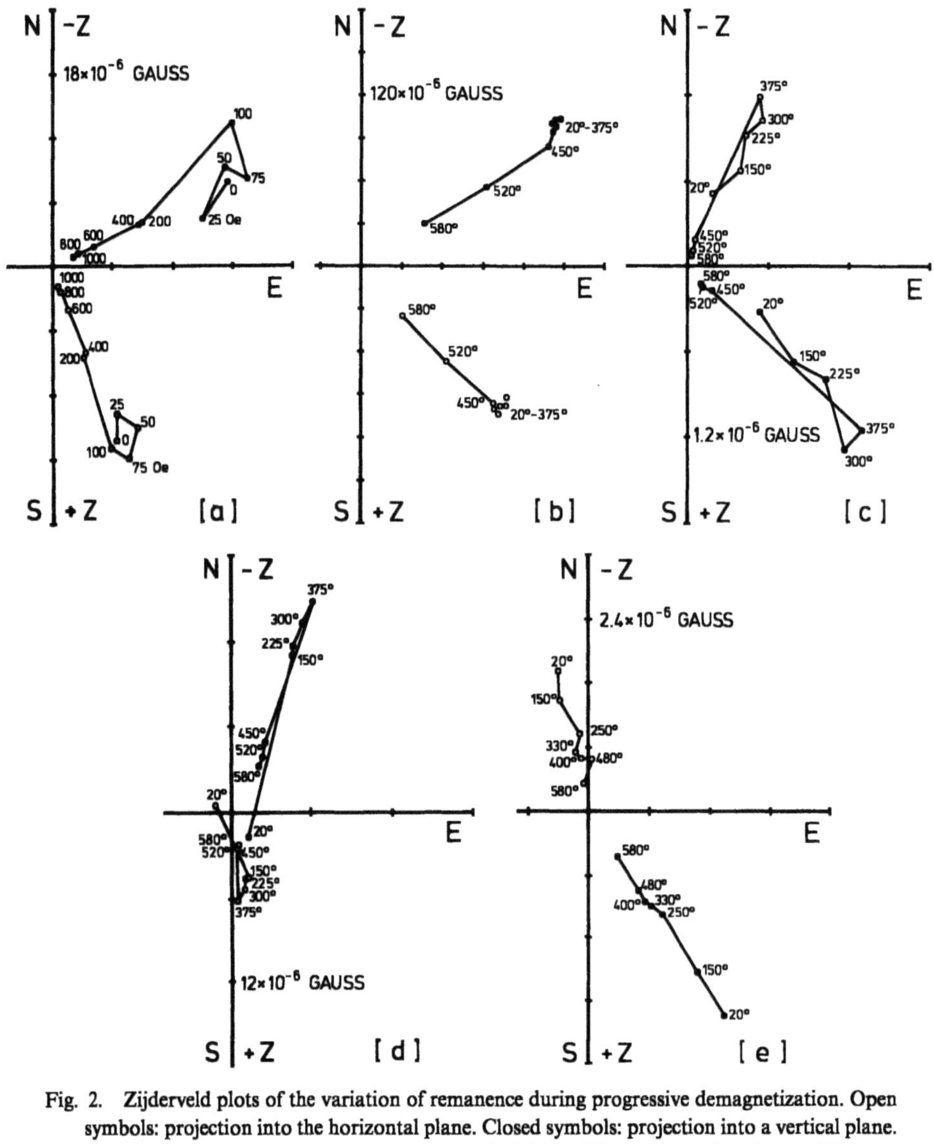

Fig. 2. Zijderveld plots of the variation of remanence during progressive demagnetization. Open symbols: projection into the horizontal plane. Closed symbols: projection into a vertical plane. a) AF demagnetization of a Lower Tertiary ignimbritic tuff from an area Southeast of Kerman (Southern Lut Block). The NRM is of reversed polarity and mainly monocomponent. b) Thermal demagnetization of a Lower Tertiary tuffaceous sandstone from the volcanic province south of Yazd. The NRM is of reversed polarity and mainly monocomponent. c) Thermal demagnetization of a Lower Cretaceous sandstone from the Shirkuh mountains south-west of Yazd. The NRM is of normal polarity. Superimposed on a stable normal component is an almost oppositely directed softer reversed component, with blocking temperatures below 400°C. d) Thermal demagnetization of a Jurassic sandstone from an area south-west of Bafq. The NRM is of normal polarity and low intensity. After the removal of a softer normal component with maximum blocking temperatures of 375°C, a stable reversed component could be isolated. e) Thermal demagnetization of a Devonian red sandstone from an area north of Kerman. NRM is of normal polarity and mainly monocomponent.

rocks were subjected to progressive thermal demagnetization. In the same way as for the AF demagnetization the proper temperature treatment for the isolation of the most stable remanence component was selected. In general there was little or no evidence for the presence of goethite (extreme high stability in AF fields, extreme low stability during thermal cleaning below 100°C) in the sandstones and limestones. This may be due to the arid climate in the sampling area. In many cases the red sandstones revealed better grouping after additional thermal cleaning with positive fold tests which could not be achieved after AF cleaning alone.

Figure 2 shows some examples of AF and thermal demagnetization of rocks of different ages, illustrated by Zijderveld diagams (ZIJDERVELD, 1967). For details see caption of Fig. 2. The pole positions of the sites selected for Fig. 2 plot all within the α_{95} cones of confidence around the mean pole position of the corresponding age interval (see Fig. 3).

3.3 *Bedding corrections and ages of rocks*

There are only very few sites of the whole sample collection where bedding corrections are not of major importance. In general the geological situation of a given sampling area was quite evident from the geological maps and other sources of information as well as from our own observations in the field during sampling. However it must be kept in mind that the geology especially in many poorly accessible parts of Central Iran and the Lut Block is not so well known in its structural details as one would have needed for the execution of absolutely reliable bedding corrections. So some of the presently negative or not very conclusive fold tests may be due to still undetected local tectonic effects.

The correct ages of some sampled formations are of course still a matter of dispute among geologists working in Central Iran. Radiometric age determinations are only rarely available and the fossil record of some of the sedimentary sequences is often so poor that classifications like "Upper Tertiary," "Lower Mesozoic" or just "Paleozoic" are the most precise information at the moment for rocks sampled from many sites.

3.4 *Pole positions derived from the most stable remanence component*

The mean pole positions computed for the sites where a reasonable grouping ($k>3$) could be obtained for the most stable remanence component are plotted in Fig. 3 for the different formations. Poles on the northern hemisphere are shown in closed symbols, those on the southern hemisphere in open symbols. As a rule only the northern poles have been plotted. In general the pole positions cluster better after application of the bedding corrections yielding at least some sort of positive fold test. Strict fold tests after GRAHAM (1949) were only rarely possible due to lack of intensive folding of the sampled sequences.

The individual pole positions which have been excluded from the computation of the formation mean are marked by a semicircle. There are several criteria which lead to the exclusion of these data.

a) Lack of precision of the remanence direction. While the pole positions derived from sites with k less than 3 have not been plotted at all, sites with $3<k<5$ were as a rule not used for the computation of the mean pole position.

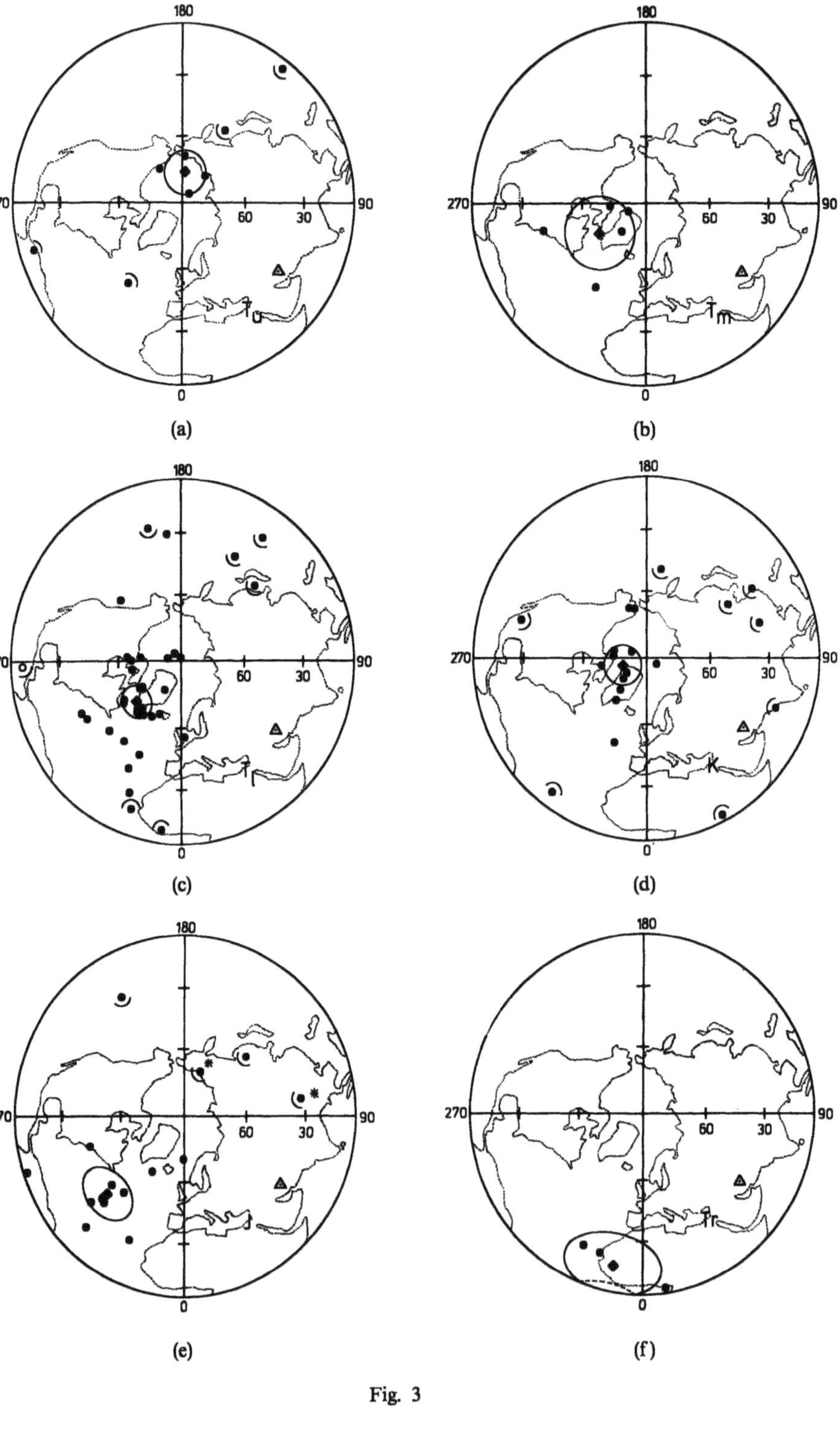

Fig. 3

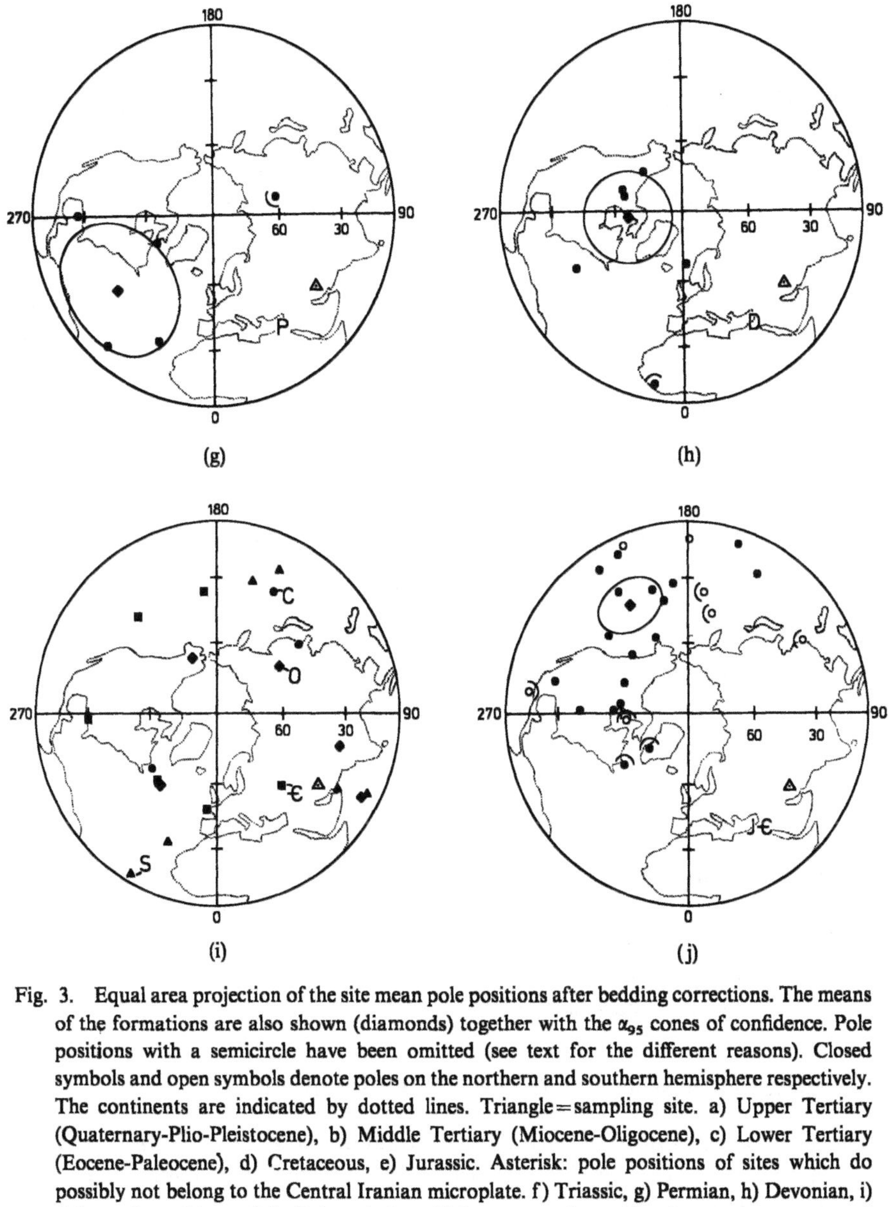

Fig. 3. Equal area projection of the site mean pole positions after bedding corrections. The means of the formations are also shown (diamonds) together with the α_{95} cones of confidence. Pole positions with a semicircle have been omitted (see text for the different reasons). Closed symbols and open symbols denote poles on the northern and southern hemisphere respectively. The continents are indicated by dotted lines. Triangle = sampling site. a) Upper Tertiary (Quaternary-Plio-Pleistocene), b) Middle Tertiary (Miocene-Oligocene), c) Lower Tertiary (Eocene-Paleocene), d) Cretaceous, e) Jurassic. Asterisk: pole positions of sites which do possibly not belong to the Central Iranian microplate. f) Triassic, g) Permian, h) Devonian, i) Other pole positions of the Paleozoic for which a mean pole position has not been computed because of too much scatter of the data. Dots: Carboniferous; Triangles: Silurian; Diamonds: Ordovician; Squares: Cambrian. j) Infracambrian.

b) Possible partial or complete magnetic overprinting. These are cases where demagnetization processes failed to show other remanence directions than the present geomagnetic (normal) field direction in the sampling area combined with a negative fold test.

c) Negative fold test. These are cases where the most stable remanence component was not along the present geomagnetic field in the sampling area and where no consistent other remanence component could be isolated and where the fold test was negative. A complete magnetic overprint of unknown age and/or inadequate bedding corrections may be the reason for the aberrant pole position.

d) Intermediate pole position. Some lava flows in regions with otherwise well grouped data revealed stable but aberrant directions which cannot be related to an abnormal tectonic situation. These directions originate possibly from intermediate field directions during a polarity transition.

e) Sites which may possibly not belong to the Central Iranian block. These sites are additionally marked with an asterisk. These are two results from the Alborz (Elburs) mountains (Jurassic melaphyrs).

The mean pole positions of the Permian and younger formations are reasonably well defined. The palaeomagnetic record of the Paleozoic is very poor. No mean pole positions have been computed for the Carboniferous, Silurian, Ordovician and Cambrian. The Devonian poles group reasonably well and provide the only information of the Paleozoic. The Infracambrian is better defined and comprises all data from the Lower Cambrian to the Young Precambrian. The time interval covered may be longer than 100 m.y. and part of the scatter may well come from a certain degree of apparent polar wandering during the Late Precambrian.

The mean pole positions are listed in Table 2 together with the statistical parameters. Table 2 shows also the number of sites of each formation and the number of sites used for the computation of the mean pole position.

3.5 Rock magnetic investigations

Four all igneous rocks the Curie temperatures (J_s/T-curves) have been determined for the identification of the major magnetic minerals. This was completed by polished section studies and measurements of hysteresis loops. In addition to the criteria set up in Section *3.4*, these data helped in many cases to explain a certain property of a rock and its behaviour during demagnetization. Lack of space does not permit to go into details. However it is planned to publish these data elsewhere.

4. Apparent Polar Wander Path (APWP) of Central Iran

The mean pole positions of the different formations (Fig. 3) are plotted in Fig. 4 in an equal area projection. Double lines are drawn between the data points to represent the APWP of Central Iran. For reference other APWP's of adjacent plates are also shown, some of them only schematically for reasons of simplicity. The following published data have been used: APWP of Eurasia (IRVING, 1977); APWP of Africa (MCELHINNY, 1973; IRVING, 1977); APWP of India (KLOOTWIJK, 1976); APWP of Arabia (MCELHINNY, 1973). Individual pole positions determined by WENSINK *et al.* (1978) and WENSINK (1979) from the Alborz (Elburs) mountains in Northern Iran, East of Tehran, are also shown (diamonds) as well as pole positions from Afghanistan (triangle) determined by KRUMSIEK (1976, 1978). All pole positions have been summarized in Table 3.

It is evident from Fig. 4 that the APWP of Central Iran is distinctly different from

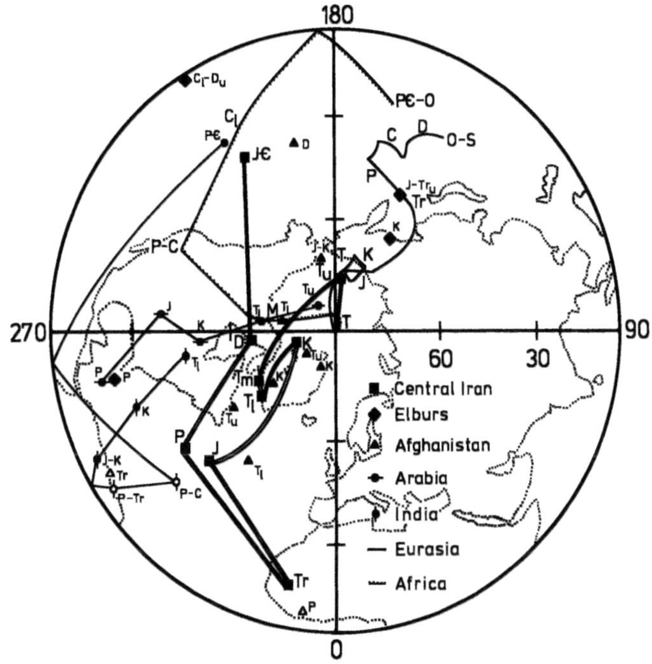

Fig. 4. Apparent polar wander paths of Central Iran (double line), Eurasia and Africa (only schematic) as well as India and Arabia. Pole positions from other investigations in Iran and Afghanistan are also included. The data are also listed in Table 3. For references see text. Continents are contoured by a dotted line.

that of Eurasia. Agreement is only given for the Uppermost Tertiary (Post Plio-Pleistocene). For prior times the APWP of Central Iran shows similar trends as the APWP's of India, Arabia, and Africa, yet the pole positions of corresponding ages are always at higher latitudes. The pole positions from Afghanistan (Krumsiek, 1976, 1978), shown as triangles in Fig. 4, are in general agreement with the APWP of Central Iran only for the time interval Cretaceous-Tertiary. A Jurassic-Cretaceous pole position is more related to the Eurasian APWP or may eventually also reflect a much younger (Upper Tertiary) magnetic overprint. A Permian pole is close to the less reliable Central Iranian Triassic pole position, while a Triassic pole is close to the Permo-Triassic pole of India. The Devonian pole position of Central Iran and Afghanistan are in poor agreement. The Afghanistan pole seems to be better related to the Devonian poles found for Africa. On the other hand the poles determined by Wensink (1979), shown as diamonds in Fig. 4, and originating from sites in the Alborz mountains, East of Tehran, in the heavily folded northern edge of Central Iran, agree poorly with our data from Central Iran. His Cretaceous and Jurassic-Triassic poles are situated on the Eurasian APWP indicating that the units from which the samples have been taken are in fact parts of Eurasia. Our own measurements on two sites from the Alborz mountains (see Fig. 3e, Jurassic) also yielded pole positions which are more related to the Eurasian APWP. They have been marked in Fig. 3e by an asterisk. It should be noted here that these Jurassic melaphyrs from the

Table 3. List of pole positions of Central Iran and adjacent microplates.

	Central Iran		N Afghan.		SE Afghan.		Central Afghan.		NW Iran		India		Arabia	
	λ°E	φ°N	λ°E	φ°N	λ°E	φ°N	λ°E	φ°N	λ°E	φ°N	λ°E	φ°N	λ° E	φ° N
T_u	174.1	76.8												
Neogene			306	54			305	80					215	82
T_m	302.6	64.1												
T_1	310.6	62.0			325	46	260	75			279	46	264	69
K	287.6	78.5	338	79			307	66	147.3	60.8	290	27	275	51
K–J			192	71							297	7		
J	314.4	38.0											265	39
J–T_r									152.6	48.7				
T_r	348.8	15.2			301	−9								
T_r–P											304	−7		
P	306.8	34.7					353	−7	281.8	22.5			282	18
P–C											312	−26		
C														
C–D									212.2	0.2				
D	276.6	66.0					193	37						
S														
O														
O–Є													323	37
C														
J-Є	209.0	35.5									212	28		
Reference	This paper		KRUMSIEK (1978)		KRUMSIEK (1978)		KRUMSIEK (1978)		WENSINK (1979)		KLOOTWIJK (1976)		MCELHINNY (1973)	

Alborz mountains are in a tectonically complicated structure and bedding corrections of nearly 180° had to be executed. The Permian pole from the Alborz mountains agrees well with that for Arabia. However the Devonian pole is significantly apart from the Central Iranian pole position and is closer to the Devonian of Africa and Afghanistan.

It must be noted in this context that there are considerable discrepancies between the pole positions determined for Central Iran and the immediately neighbouring areas like the Alborz mountains and Afghanistan. It is our opinion that this may result in part from insufficient palaeomagnetic data (especially from the Lower Mesozoic and entire Paleozoic) and/or geological control on the tectonic history of the units. Another explanation could be that the entire region is composed of different smaller microplates with individual drift histories.

In the following we will concentrate on the discussion of the APWP of Central Iran and its relation mainly to the Eurasian, to some extent also to the African and Indian plates.

5. Palaeoremanence Directions

They were computed for reference points in Eurasia (Turan plate, 56°E, 38°N), India (80°E, 20°N), Arabia (45°E, 25°N), Central Iran (56°E, 33°N), Northwest Iran (52°E, 36°N), and Northern, Central, and South-east Afghanistan (65°E, 36°N; 65°E, 33°N; 65°E, 30°N) using the corresponding APWP's and pole positions listed in Table 3. The palaeodeclinations and inclinations of the reference points are listed in Table 4 for the different ages together with the palaeolatitude, which was computed from the palaeoinclination assuming an axial geocentric dipole field. Negative palaeolatitudes in Table 4 indicate a palaeoposition of the reference point on the lower hemisphere.

5.1 Palaeolatitudes

The palaeolatitudes of the reference points in India and Arabia show the well known drift of India from the lower hemisphere (negative palaeolatitudes) towards the Eurasian landmass since Mesozoic time as well as the northward movement of Africa and Arabia due to the displacement of that plate and its anticlockwise rotation. The reference point on the Eurasian plate (Turan) maintained its position between 30° and 40°N almost throughout the entire Post Triassic time, after it has moved northward from about 10°N to 35°N between Carboniferous and Lower Jurassic time. Table 4 shows that the reference point in Central Iran was in an intermediate latitude position between Eurasia and the Gondwana landmasses since the Late Paleozoic. The palaeolatitudes computed for the Alborz mountains in Northwest Iran from the data of Wensink (1979) and for Afghanistan using the data of Krumsiek (1978) give the same general result. The palaeolatitude situation of Northwest Iran, Central Iran, and Afghanistan for the Middle and Lower Paleozoic is still ill defined. The Northwest Iranian and the Afghanistan data point more to a southern position than the reference point on the Turan plate, while our Devonian palaeolatitude of the reference point in Central Iran agrees with that on the Turan plate. The palaeolatitudes of Central Iran and India agree for the Infracambrian. This was the first palaeomagnetic evidence for the concept of Central Iran being a former

Table 4. Predicted palaeoremanence directions and palaeolatitudes φ of reference

	Central Iran			N. Afghanistan			SE. Afghanistan			Central Afghan.		
	D°	I°	φ° N	D°	I°	φ° N	D°	I°	φ° N	D°	I°	φ° N
T, Upper	13.0	44.4	26.1									
Neogene				328.0	26.8	14.2				350.2	46.3	27.6
T, Middle	334.7	36.3	20.1									
T. Lower	330.8	39.1	22.1				315.0	27.8	14.8	356.0	33.7	18.5
K	350.0	43.6	25.4	346.4	55.3	35.8				337.6	35.6	19.7
K–J				16.5	40.9	23.4						
J	307.9	22.4	11.7									
J–T_r												
T_r	271.2	45.8	27.2				279.5	−53.3	−33.8			
T_r–P												
P	308.8	9.5	4.8							285.9	21.3	11.0
P–C												
C												
C–D												
D	344.2	26.2	13.8									
S												
O												
O–€												
€												
J€	22.8	−31.4	−17.0									
Ref. site	56°E, 33°N			65°E, 36°N			65°E, 30°N			65°E, 33°N		

They were computed from the mean pole positions (Table 3) assuming a geocentric axial dipole field. The last two rows give information about the relative rotational movement between Central Iran and the Southern margin of Eurasia (Turan plate) as well as the relative displacement of Central

part of Gondwanaland (BECKER *et al.*, 1973).

The present latitude difference between the reference points on the Turan plate and in Central Iran is 5°. Additional palaeolatitude differences can be interpreted in terms of a North-South displacement of both units in the past. These data have been listed in Table 4. The largest southward displacement of Central Iran with respect to the Turan plate occurred in the Jurassic and amounted to about 16°. Based on our palaeomagnetic data from Central Iran there was no or no significant relative North-South displacement between Central Iran and the Turan plate throughout the Lower and Middle Paleozoic. Both units were at more or less the same palaeolatitude. After a slight trend towards the South during the Permian, Central Iran moved again northward between the Permian and the Triassic. Between the Triassic and the Jurassic Central Iran moved to the South into a position relative to the Turan plate, which has already been described above. Since the Jurassic, Central Iran had a more or less steady tendency to move northward towards the Turan plate.

As can be seen in Table 4, the palaeolatitude positions of Northwest Iran and Afghanistan show sometimes an agreement but sometimes also marked differences with our data from Central Iran. More palaeomagnetic measurements are needed to find out sites on various microplates adjacent to Central Iran for different ages.

NW. Iran			India			Arabia			Turan Plate			Angle of rotation	North-South displacement
D°	I°	φ°N	D°	I°	φ°N	D°	I°	φ°N	D°	I°	φ°N	(°)	(°)
									3.7	53.5	34.0	9.3 cl. w.	2.9 S.
						1.5	31.6	17.1					
									11.7	49.9	30.7	36.9 c.cl.w.	5.6 S.
			345.9	−38.6	−21.8	346.8	16.0	8.2	15.1	48.8	29.7	44.3 c.cl.w.	2.6 S.
33.6	47.5	28.5	327.2	−54.2	−34.7	331.2	−4.4	−2.2	22.2	51.6	32.2	32.2 c.cl.w.	1.8 S.
			302.9	−63.2	−44.7								
						328.7	−29.6	−15.9	16.0	52.2	32.8	68.1 c.cl.w.	16.1 S.
43.7	36.1	20.1											
									46.8	44.2	26.0	135.6 c.cl.w.	6.2 N.
			280.7	−63.8	−45.5								
312.8	−27.6	−14.9				302.0	−35.8	−19.8	44.7	29.5	15.8	95.9 c.cl.w.	6.0 S.
			252.6	−61.0	−42.1								
									52.3	20.8	10.7		
30.8	−66.9	−49.4											
									60.8	22.3	11.6	76.6 c.cl.w.	7.2 N.
									65.7	27.6	14.6		
						302.2	37.2	20.8					
									54.0	−44.9	−26.5		
			45.6	−40.1	−23.2								
52°E, 36°N			80°E, 20°N			45°E, 25°N			56°E, 38°N				

Iran in the North-South direction. Clockwise=cl.w.; counterclockwise=c.cl.w.; 2.9° South means a displacement of Central Iran of 2.9° to the South with respect to Southern Eurasia.

whether these differences are significant and represent movements of different crustal units in this region.

5.2 Rotational movements of Central Iran with respect to Southern Eurasia

While the palaeolatitude differences between the Turan plate and Central Iran are much less than those between Southern Eurasia and India, the palaeodeclinations of India and Central Iran are quite similar (see Table 4) and indicate large amounts of counterclockwise rotations of these plates with respect to Southern Eurasia since the Late Paleozoic. The angles of total counterclockwise rotation between Central Iran and the Turan plate are listed in Table 4 for the different ages. The largest value of about 135° is observed for the Triassic. Between the Middle Paleozoic and the Lower Mesozoic there was a clockwise rotation of about 60° between Central Iran and Southern Eurasia. This was followed by a counterclockwise rotation of about 65° between the Triassic and the Jurassic, a further 35° counterclockwise rotation between the Jurassic and the Middle Tertiary and finally an about 35° counterclockwise rotation since Middle Tertiary times.

6. Discussion and Conclusions

Based on other reconstructions on the basis of palaeomagnetic data of the Gondwana landmasses and continents around the present Indian ocean (e.g. BRIDEN *et al.*, 1974; IRVING, 1977) a reconstruction of the position of Central Iran was attempted with the presently available palaeomagnetic data. This is shown in Fig. 5. The most reliable part as far as Central Iran is concerned is the Post Triassic time. In this time interval there is also reasonable agreement with the data from Afghanistan, although both units including also Northwest Iran may not have moved always in coordination. For earlier times there are still considerable uncertainties due to the lack of sufficiently reliable palaeomagnetic data. Furthermore the problem of the indeterminacy of palaeolongitude becomes increasingly important for older times.

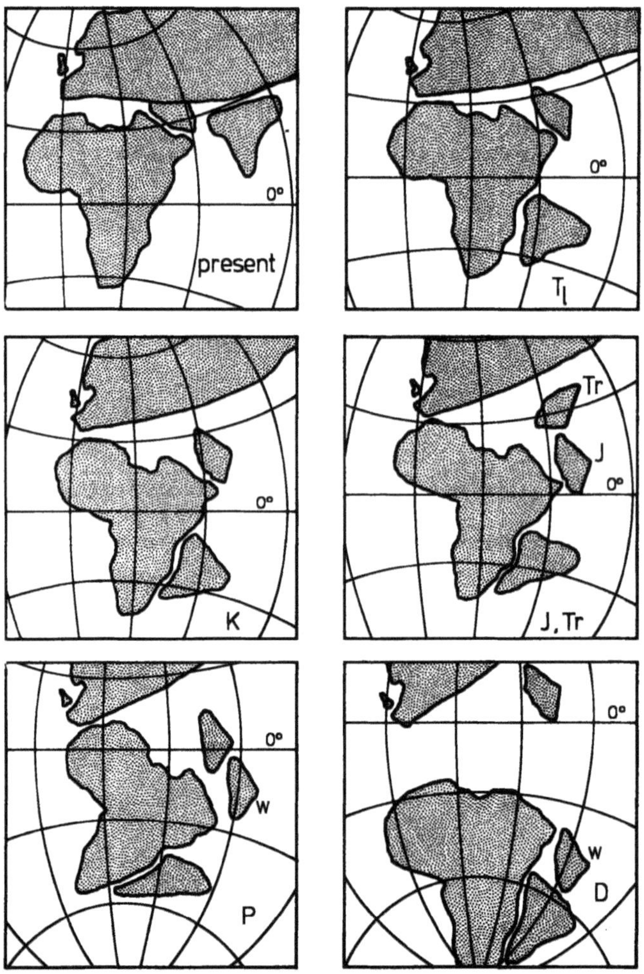

Fig. 5. Tentative reconstruction of the position of Central Iran with regard to Eurasia, Africa, and India since the Paleozoic. For the Permian and the Devonian our reconstruction as well as that of Wensink (W) is shown. Lambert projection, grid intervals: 30°.

By including also palaeogeographical data into the considerations, the following general scheme for the drift of the Central Iranian microplate is obtained.

Our palaeomagnetic data suggest that Central Iran was an integral part of Gondwana during Precambrian, Infracambrian, and possibly also Cambrian time. Gondwana underwent uplift for a least 100 m.y. at the end of the Precambrian delivering large masses of terrigeneous material into continental rift valleys which mark the later lines of separation. The Infracambrian rhyolites are the expression of the magmatism corresponding to the crustal thinning and incipient rifting. In Paleozoic time a sedimentary gap between the Ordovician and the Permian is typical for Iran as well as for Arabia, India, and the Tarim Basin (ANONYMOUS, 1976, p. 8). This gap explains the relatively small number of sampling localities for Paleozoic rocks in Central Iran. In Upper Paleozoic time Central Iran drifted across the equator. The tropical climate is reflected by high alumina clays in Permian beds near Abadeh containing a recently discovered Linopteris-Walchia flora with possible Gondwana connections (Hagemann, private communication). Upper Permian traps are known throughout Central Iran. The similarity between Central Iran, Central Afghanistan (LAPPARENT *et al.*, 1970) and the Pamir (KARAPETOV *et al.*, 1975) in Permo-Carboniferous time is striking. The idea developed here involves an ocean north of Central Iran and its eastern continuation in Paleozoic time. This Paleotethys (LAPPARENT *et al.*, 1970) was closed in Permo-Carboniferous time leaving the Southern Caspian Depression and Ophiolite lenses farther east (Mashhad fault zone) as remnants. The closure implies the stepwise collision of parts of the Central Domain with Eurasia in Hercynian and Early Kimmerian time (North Pamir, Turan). Central Iran had a much earlier collision with Eurasia than India. Due to its collision, northern parts of Gondwana lost their platform character during the Triassic. The fragmentation was followed by trap like effusions in Lower and Middle Jurassic time (Alborz). Sedimentation changed abruptly from shelf carbonates to continental coal bearing sandstones. During Upper Triassic time a land bridge has been established between Central Asia (Turan) and Central Iran (exchange of coal forming flora, e.g. Dictiophyllum).

Central Iran was detached from Gondwana in Paleozoic time. This implies the formation of an ocean south of Central Iran. We regard the margin between Central Iran and that Neotethys as passive. Oceanic basalts, sheeted diabases, layered intrusions and ophiolites, formed at midocean ridges, are covered by Late Triassic to Middle Cretaceous sediments e.g. at Neyriz (Atlantic flysch, FÖRSTER, 1976; outer subbelt, STÖCKLIN, 1977).

The evolution since Upper Triassic time in Central Iran is characterized by Early Jurassic grabens, Jurassic folding-thrusting-metamorphism-granite intrusion, and Late Jurassic/Early Cretaceous red beds and trachybasalts (FÖRSTER, 1976) and Middle Cretaceous shelf carbonates. The palaeogeographic connection during Lower Cretaceous time with Afghanistan and South Tadshikistan has been demonstrated (FÖRSTER, 1976). Our palaeomagnetic data indicated a much closer neighbourhood to Turan and a much larger gap (ocean?) between Central Iran and the more southern landmasses (Arabia, East Africa, India).

During the Cretaceous the nature of the southern margin of Central Iran changed from Atlantic to active Andean type: flysch type silt-sandstone, pillowed diabase, pelagic Late Cretaceous/Paleocene limestone, exotic blocks (olistolites) (inner subbelt, STÖCKLIN, 1977). Arabia advanced and pushed oceanic crust of the Neotethys in a trench paralleling the Zagros thrust northeastward below Central Iran.

The palaeolatitude position of Central Iran in the Tertiary and Quaternary is well documented by measurements in the volcanic series piled up in the Central Iranian Volcanic Belt and in the Lut Volcanic Zone. There are some transgressions of the Neotethys into the Central Iranian Block (Baft-Nain, Birjand, Sabzevar, see Fig. 1). According to LENSCH *et al.* (1979) the ophiolites north of Sabzevar might represent an independent and older oceanic basin.

The existence of a narrow ocean between Turan and Central Iran (Sabzevar) is demonstrated by our data at the end of Mesozoic time (difference in palaeolatitude). Remnants of this basin are covered with Eocene and Neogene volcanics (Sabzevar Volcanic Zone). This volcanism is likely to be related to incipient rifting, subduction and basin and range tectonics. The closing of oceanic basins is connected with the formation of high pressure metamorphic assemblages in the supposed subduction zones at Sabzevar Neyriz (VIALON *et al.*, 1972).

During the Tertiary Central Iran held a position between two huge jaws (Turan and Arabia). This is reflected by Early Tertiary folding which affected also the Eocene volcanics of the Central Iranian Volcanic Belt. Central Iran emerged after that folding and acted as a relatively stable block during the younger folding phase (Pliocene).

We thank the Deutsche Forschungsgemeinschaft for the generous financial support of several sampling trips to Iran between 1972 and 1977. Special thanks are due to the Geological Survey of Iran for the help and advice during the field work.

REFERENCES

ANONYMOUS (Compilation Group of the Geological Map of China), *An Outline of the Geology of China*, 22 pp., Chinese Academy of Geological Sciences, Peking, 1976.

BECKER, H., H. FÖRSTER, and H. SOFFEL, Central Iran, a former part of Gondwanaland? Palaeomagnetic evidence from Infra Cambrian rocks and iron Ores of the Bafq area, Central Iran, *Zeitschr. Geophys.*, **39**, 953–963, 1973.

BRIDEN, J., G. DREWRY, and G. SMITH, Phanerozoic equal area world maps, *J. Geol.*, **82**, 555–574, 1974.

BRUNN, J. H., Über die Entstehung gefalteter Ketten: Kollisions-Tektonik und induzierte Bögen, *Z. dt. Geol. Ges.*, **127**, 323–335, 1976.

CRAWFORD, A. R. and W. COMPSTON, The Age of the Vindhyan System of Peninsular India, *Quart. J. Geol. Soc. Lond.*, **125**, 351–372, 1970.

CRAWFORD, A. R., Iran, continental drift and plate tectonics, 24th Intern. Geol. Congr. Sect. 3, pp. 71–80, Montreal, 1972.

DEWEY, J. F., W. C. PITMAN III, W. B. F. RYAN, and J. BONNIN, Plate tectonics and the evolution of the Alpine system, *Bull. Geol. Soc. Am.*, **84**, 3137–3180, 1973.

FÖRSTER, H., Associations of volcanic rocks in the mountains south of Sabzevar (NE Iran), 23rd Intern. Geol. Congr. Sect. 2, pp. 197–212, Prague, 1968.

FÖRSTER, H., Continental Drift in Iran in relation to the Afar Structures, in *Afar between Continental and Oceanic Rifting*, Vol. 2, edited by A. Pilger and A. Rösler, pp. 182–190, 1976.

GRAHAM, J. W., The stability and significance of magnetism in sedimentary rocks, *J. Geophys. Res.*, **54**, 131–167, 1949.

HAHN, G. and H. D. PFLUG, Ein neuer Medusen-Fund aus dem Jung-Präkambrium von Zentral-Iran, *Senckenbergiana laet.*, Frankfurt, 1980 (in press).

IRVING, E., Drift of the major continental blocks since the Devonian, *Nature*, **270**, 304–309, 1977.

KARAPETOV, S. S., I. I. SONIN, and V. E. KHAIN, On some of the most important characteristics of the structure and development of the Afghan-Pamir segment of the Alpine folded belt of Eurasia, *Vestnik Mosk. Univ.*, **3**, 28–46, 1975 (in russian).

KASHFI, M. S., Plate tectonics and structural evolution of the Zagros geosyncline, southwestern Iran, *Geol. Soc. Am. Bull.*, **87**, 1486–1490, 1976.

KLOOTWIJK, C. T., The drift of the Indian subcontinent; an interpretation of recent palaeomagnetic data, *Geol. Rdsch.*, **65**, 885–909, 1976.

KRUMSIEK, K., Zur Bewegung der Iranisch-Afghanischen Platte, *Geol. Rdsch.*, **65**, 909–929, 1976.

KRUMSIEK, K., Polwanderungskurven afghanischer Krustenschollen, Konstruktion und Deutung, Habil. Thesis, Univers. Bonn, 1978.

LAPPARENT, A. F. DE, H. TERMIER, and G. TERMIER, Sur la stratigraphie et la paléobiologie de la série permo-carbonifère du Dacht-e-Newar (Afghanistan), *Bull. Soc. Géol. France*, **12**, 565–572, 1970.

LENSH, G., A. MIHM, and N. ALAVI-TEHRANI, Major element geochemistry of the ophiolites north of Sabzevar (Iran), *N. Jb. Geol. Paläont. Mh.*, **7**, 415–447, 1979.

MCELHINNY, W. M., *Palaeomagnetism and Plate Tectonics*, pp. 185–234, The University Press, Cambridge, 1973.

SOFFEL, H., H. FÖRSTER, and H. BECKER, Preliminary polar wander path of Central Iran, *J. Geophys.*, **41**, 541–543, 1975.

STÖCKLIN, J., Structural history and tectonics of Iran—a review, *Am. Assoc. Petr. Geol. Bull.*, **52**, 1229–1258, 1968.

STÖCKLIN, J., Stratigraphic Lexicon of Iran, part I: Central, North and East Iran, Geol. Survey of Iran, Rep. No. 18, 338 pp., 1971.

STÖCKLIN, J., Possible ancient continental margins in Iran, in *The Geology of Continental Margins*, edited by C. A. Burke and C. L. Drake, pp. 873–887, Springer, New York, 1974.

STÖCKLIN, J., Structural correlation of the Alpine ranges between Iran and Central Asia, *Mém. h. sér. Soc. Géol. France*, **8**, 333–353, 1977.

TAKIN, M., Iranian geology and continental drift in the Middle East, *Nature*, **235**, 147–150, 1972.

VIALON, P., A. HOUCHMAND-ZADEH, and M. SABZEHI, Proposition d'un modèle de l'évolution pétro-structurale de quelques montagnes iraniennes, comme une conséquence de la tectonique de plaques, 24th Intern. Geol. Congr. sect. 3, pp. 196–208, Montreal, 1972.

WENSINK, H., J. D. A ZIJDERVELD, and J. C. VAREKAMP, Palaeomagnetism and ore mineralogy of some basalts of the Geirud formation of Late Devonian-Early Carboniferous age from the Southern Alborz, Iran, *Earth Planet. Sci. Lett.*, **41**, 441–450, 1978.

WENSINK, H., The implications of some palaeomagnetic data from Iran for its structural history, *Geol. en Mijnbouw*, **58**, 175–185, 1979.

ZIJDERVELD, J. A. D., A. C. Demagnetization of Rocks: Analysis of Results, in Developments in Solid Earth Geophysics, Vol. 3, *Methods in Palaeomagnetism*, edited by D. W. Collinson, K. M. Creer, S. K. Runcorn, pp. 254–286, Elsevier Publ. Comp., Amsterdam-London-New York, 1967.

Reversals and Excursions of the Geomagnetic Field as Defined by Palaeomagnetic Data from Upper Palaeozoic-Lower Mesozoic Sediments and Igneous Rocks from Argentina

Daniel A. VALENCIO

Departamento de Ciencias Geológicas, Universidad de Buenos Aires, Ciudad Universitaria, Pabellón, Buenos Aires, Argentina

(Received December 13, 1979)

Palaeomagnetic data from samples collected in rock sequences of sediments and igneous rocks of Late Palaeozoic and Triassic age exposed in northwestern Argentina are summarized. On basis of these data a composite magnetic stratigraphy for the Late Carboniferous, Permian and Early and Middle Triassic was prepared.

They record two normal polarity events at the Late Carboniferous and Middle Permian which are correlated with the Debal'tseva Zone and the Quebrada del Pimiento Normal Event, respectively.

They also record short-lived normal polarity events and geomagnetic excursions at the Late Permian-Early Triassic, which are correlated with the Illawarra Zone, and after them a long period of predominant normal polarity. This is correlated with the normal quiet interval reported within the Triassic; there are many geomagnetic excursions within this interval which are being investigated.

1. Introduction

In northwestern Argentina, large exposures of continental sedimentary deposits and igneous rocks have been assigned a Late Palaeozoic to Early Mesozoic age. Palaeomagnetic data for three sequences of these rocks are used to define the chronology of the reversals and excursions of the geomagnetic field in Late Palaeozoic-Early Mesozoic times.

2. Available Data

2.1 Geological evidence

In the northwestern part of Argentina red beds and igneous rocks formed in Late Palaeozoic-Early Mesozoic time are exposed; these rocks have been recently classified in two subgroups within the Paganzo Group and one formation (AZCUY *et al.*, 1978). Named in chronological order from the base to the top, they are: Section I (Paganzo I) and Section II (Paganzo II) of the Paganzo Group and the Amaná-Talampaya Formation (Paganzo III). A Permocarboniferous age has been assigned for the rocks included in

Section II of the Paganzo Group; a Late Permian to Triassic age was suggested for the red beds included in the Amaná-Talampaya Formation.

The sequences of Paganzo red beds exposed near Paganzo village (30.3°S, 67.7°W, site A) and Mina Las Mellizas (30°S, 67°W, site B), Province of La Rioja, have been assigned to La Colina Formation, which is correlated with Section II of the Paganzo Group. Two igneous units are intercalated in La Colina red beds at these sites. A basaltic sill (266 ± 7 Myr, VALENCIO *et al.*, 1977) is exposed close to the base of La Colina rocks sequence in site A; a basaltic lava flow (295 ± 5 Myr, THOMPSON and MITCHELL, 1972) is exposed in the middle of La Colina rocks sequence in site B. The sequences of red beds which rest unconformably on La Colina rocks in both sites (A and B) have been assigned to the Amaná-Talampaya Formation.

Also in the northwestern part of Argentina (Province of Catamarca) red beds and sediments formed in Late Palaeozoic and Early Mesozoic time are exposed; these rocks have been classified in three formations. Named in chronological order, from the eldest to the youngest, they are: De la Cuesta Formation (Permian), Quebrada la Coipa Formation (Triassic) and Del Crestón Formation (Triassic) (TETZLAFF, 1978). A sequence of about 1,000 m of sediments assigned to these formations is exposed close to Río Colorado (28.3°S, 68°W, site C).

2.2 *Palaeomagnetic data*

One hundred and eighty two hand samples were collected in a rock sequence of about 620 m of red beds an igneous rocks assigned to La Colina and Amaná-Talampaya Formations exposed near Paganzo village (site A) (VALENCIO *et al.*, 1977). After thermal and A.C. cleaning the majority of La Colina samples show reversed polarity and yield a palaeomagnetic pole at 78°S, 249°E (SAP_7, $\alpha_{95} = 3°$). They also record a normal polarity event. The stable remanent magnetization isolated in samples from the Amaná-Talampaya Formation is predominantly reversed and reveals three normal events and three geomagnetic excursions. The samples of reversed polarity yield a palaeomagnetic pole at 75°S, 285°E ($SAPTr_2$, $\alpha_{95} = 13°$). Figure 1 shows the magnetic stratigraphy for La Colina and Amaná-Talampaya rocks from site A; it consists of logs of sample-mean values of stable declination and inclination plotted alongside the log of the angle between the virtual geomagnetic pole (VGP) for each hand sample and the palaeomagnetic pole through these formations.

One hundred and one hand samples were collected in a rock sequence of about 300 m of red beds and igneous rocks, assigned to La Colina and Amaná-Talampaya Formations, exposed close to Mina Las Mellizas (site B) (SINITO *et al.*, 1979). After thermal, A.C. and chemical cleaning the majority of La Colina samples showed reversed polarity and yielded a palaeomagnetic pole at 49°S, 343°E (SAC_4, $\alpha_{95} = 5°$). These samples also recorded a normal polarity event. After cleaning the samples from the Amaná-Talampaya Formation revealed frequent reversals of declination and inclination which were associated with short-lived polarity events or geomagnetic excursions. The magnetic stratigraphy for La Colina red beds and igneous rocks from site B is shown in Fig. 1.

Four hundred and seventeen samples were collected in a sequence of about 940 m of red beds and sediments, assigned to De la Cuesta, Quebrada la Coipa and Del Crestón

Formations, exposed near Río Colorado (site C). The palaeomagnetic study of these samples is not finished yet; in this paper the preliminary results obtained up to day (TETZLAFF, 1978; SELLÉS MARTÍNEZ, 1979; GLEIZER, 1979) are reported. After thermal cleaning the De la Cuesta samples show predominant reversed polarity; they also record two normal polarity events and three geomagnetic excursions. The reversed and normal samples yield a palaeomagnetic pole at 83.0°S, 113.6°E (SAP_{10}, $N=33$, $\alpha_{95}=7.7°$).

After thermal and chemical cleaning the samples from Quebrada la Coipa Formation show predominant normal remanent magnetization; reversed remanence was only found in the lowest samples of the sequence. They also record reversals of declination and inclination which, at present, are associated with seventeen geomagnetic excursions. The normal polarity samples yield two palaeomagnetic poles for the Lower (44°S, 196°E, $SATr_8$, $N=25$, $\alpha_{95}=6.7°$) and the Upper Members (47°S, 192°E, $SATr_9$, $N=31$, $\alpha_{95}=5.2°$) of this formation.

After thermal and chemical treatment all the samples from Del Crestón Formation showed normal polarity and yielded a palaeomagnetic pole at 52°S, 189°E ($SATr_{10}$, $N=58$, $\alpha_{95}=4.2°$). The magnetic stratigraphy for De la Cuesta red beds, Quebrada la Coipa red beds and Del Crestón sediments from site C is shown in Fig. 1.

3. Interpretation

The position of the palaeomagnetic pole for La Colina Formation exposures at site B (SAC_4, Mina Las Mellizas) is close to the positions of the Late Carboniferous poles for South America. This is consistent with the K-Ar age for the basaltic lava flow (295 ± 5 Myr) situated in the middle of the sequence studied at this site. Therefore, it seems reasonable to assign a Late Carboniferous age for La Colina rocks exposed at site B; the short polarity event recorded at the base of the sequence should be, then, correlated with the latest of the normal polarity events which define the Debal'tseva Zone (Fig. 1, SINITO *et al.*, 1979).

The mean K-Ar age (266 ± 7 Myr) for the basaltic sill exposed at the base of the sequence La Colina red beds near Paganzo village (site A), is indicative of a Middle Permian age. Therefore the age of the stratigraphically lowest red beds in this sequence should be older than Middle Permian. The position of the palaeomagnetic pole for La Colina rocks exposed at this site (SAP_7) is close to the positions of the Middle Permian poles for South America. Therefore, it is reasonable to accept a late Early Permian-early Middle Permian age for the lowest red beds of the sequence. The short-lived polarity event recorded in the middle of the sequence should then be correlated with the Middle Permian Quebrada del Pimiento Normal Magnetic Event (263 ± 5 Myr) (Fig. 1, VALENCIO *et al.*, 1977).

The frequent reversals of polarity and excursions of the geomagnetic field suggest an Illawarra Zone age for the Amaná-Talampaya red beds which rest unconformably on La Colina rocks at site A (Fig. 1). The position of the palaeomagnetic pole for those red beds ($SAPTr_2$) is consistent with this.

This reasoning can be also used to correlate the sequence of De la Cuesta red beds and the stratigraphically lowest sediments of the sequence of Quebrada la Coipa rocks

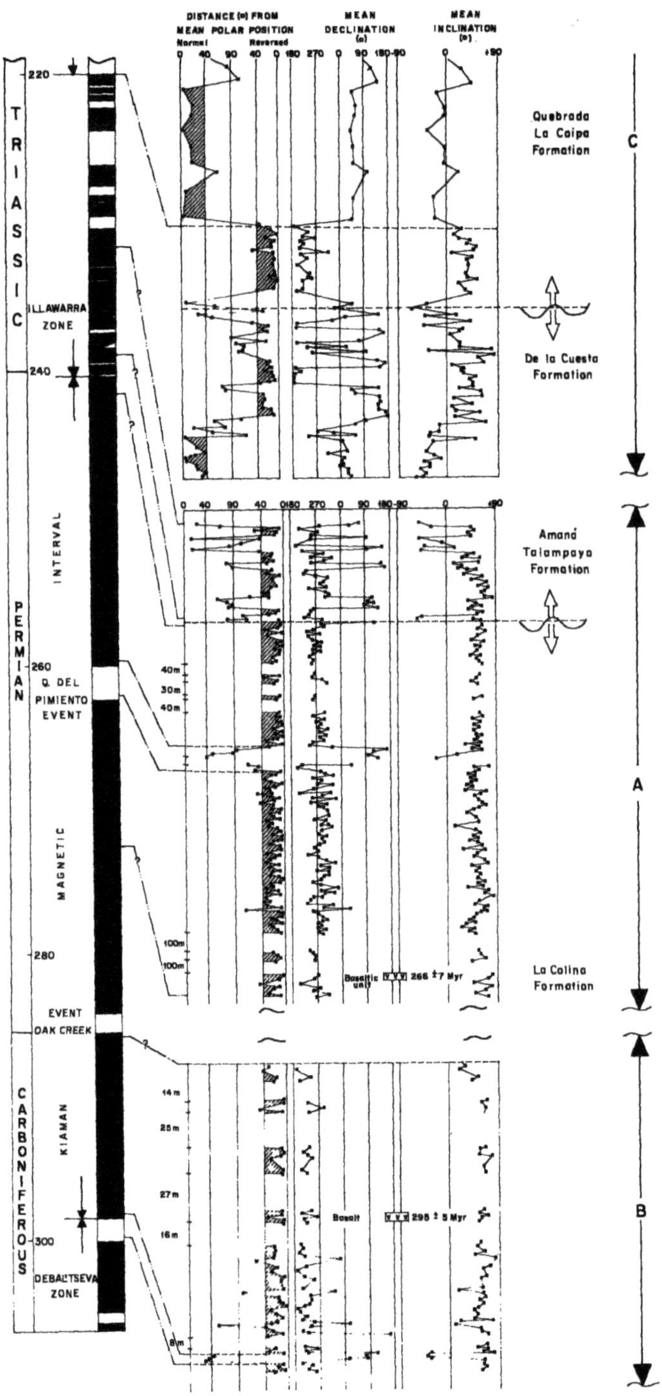

(a)

Fig. 1. Composite magnetic stratigraphy for the Late Palaeozoic-Triassic interval; it was integrated (a and b) on basis of palaeomagnetic data for sequences of red beds, sediments and igneous rocks from northwestern Argentina. An overlapping may be present between the

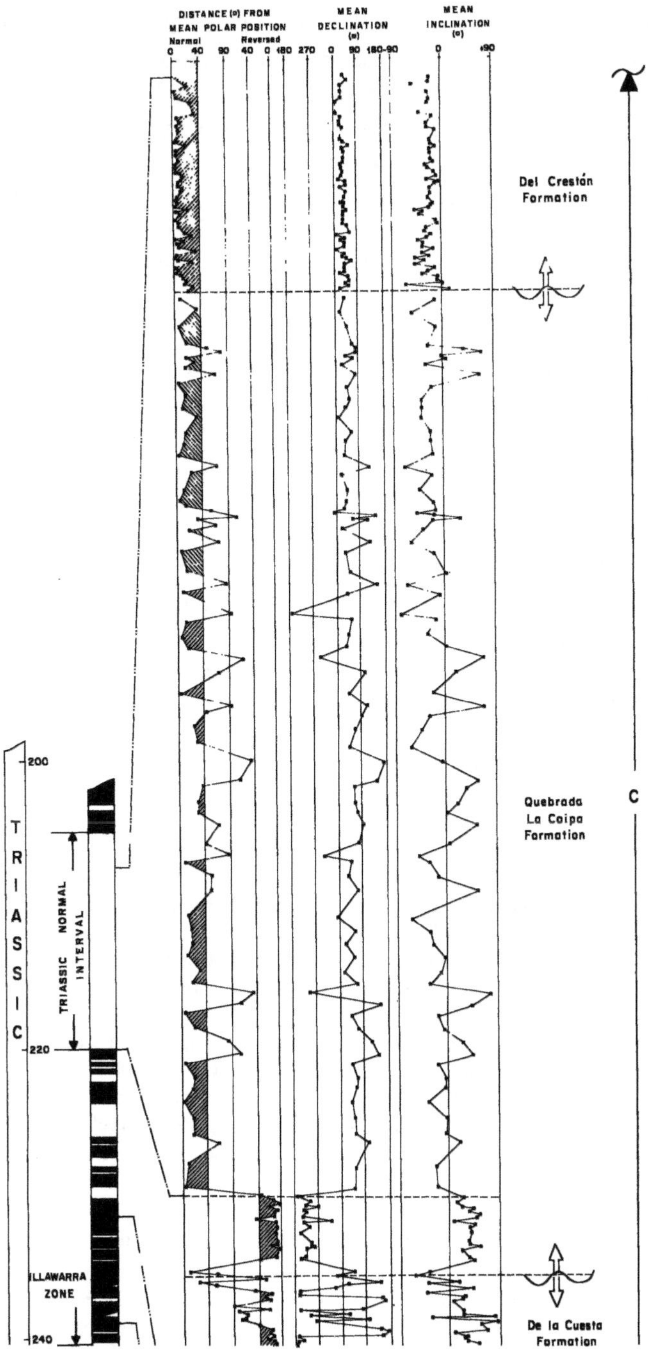

(b)

sequence of Amaná-Talampaya red beds (site A) and the sequence of De la Cuesta and Quebrada la Coipa red beds (site C). ▭, normal polarity; ■, reversed polarity of the geomagnetic field. The reversals in between the Illawarra Zone and the Triassic Normal Interval are schematic.

exposed at site C (Río Colorado), with the Illawarra Zone (Fig. 1). We should note, however, that the palaeomagnetic data do not allow us to know if there is any overlapping between the sequence of Amaná-Talampaya red beds and the sequence of De la Cuesta and Quebrada la Coipa red beds, assigned to the Illawarra Zone.

The predominant normal remanence found in the samples from the higher stratigraphic levels of the sequence of Quebrada la Coipa red beds should be correlated with the interval (220–205 Myr) of normal polarity of the geomagnetic field defined within the Triassic (IRVING and PULLAIAH, 1976) (Fig. 1). However, the pole positions for the Lower ($SATr_8$) and Upper Members ($SATr_9$) of this formation are not close to the positions of the other Triassic poles for South America. This reasoning is also valid for the normal polarity samples from the sequence of Del Crestón sediments which rest unconformably on La Coipa rocks at site C (Fig. 1). The different positions of poles $SATr_8$, $SATr_9$ and $SATr_{10}$ and the other Triassic poles for South America could be the effect of a deviation of the geomagnetic field sometime in the interval 220–205 Myr. This deviation has not been observed on the Triassic palaeomagnetic data for the other gondwanic continents.

On basis of the above discussion it was made the composite magnetic stratigraphy for the Late Carboniferous (about 295 Myr) to Triassic interval shown in Fig. 1. We should note that there are some gaps in between and within the sequences. However, the figure shows a nice picture of the behaviour of the geomagnetic field during its quiet and disturb periods within the Late Carboniferous-Triassic interval.

The author wishes to thank the Consejo Nacional de Investigaciones Científicas y Técnicas and the Universidad de Buenos Aires for the support which enabled the work described to be carried out.

REFERENCES

AZCUY, C. L., J. R. MORELLI, D. A. VALENCIO, and J. F. VILAS, Estratigrafía de la Comarca Amaná-Talampaya, *Sépt. Congr. Geológ. Argent.*, **1**, 243–256, 1978.

GLEIZER, H. M., Variaciones paleoseculares del campo magnético terrestre registradas en sedimentos del Paleozoico Superior y Mesozoico Inferior del Noroeste Argentino, Trabajo Final de Licenciatura, Universidad de Buenos Aires, 1979 (unpublished).

IRVING, E. and G. PULLAIAH, Reversals of the geomagnetic field, magnetostratigraphy, and relative magnitude of paleosecular variation in Phanerozoic, *Earth Sci. Rev.*, **12**, 35–64, 1976.

SELLÉS MARTINEZ, J., El estudio paleomagnético de sedimentos y su utilización en la correlación geológica, Trabajo Final de Licenciatura, Universidad de Buenos Aires, 1979 (unpublished).

SINITO, A. M., D. A. VALENCIO, and J. F. VILAS, Palaeomagnetism of a sequence of Upper Palaeozoic-Lower Mesozoic red beds from Argentina, *Geophys. J. R. Astr. Soc.*, **58**, 237–247, 1979.

TETZLAFF, D. M., Geología y paleomagnetismo de formaciones aflorantes en las adyacencias del Río Colorado, Departamento de Tinogasta, Provincia de Catamarca, Trabajo Final de Licenciatura, Universidad de Buenos Aires, 1978 (unpublished).

THOMPSON, R. and J. G. MITCHELL, Palaeomagnetic and radiometric evidence for the age of the lower boundary of the Kiaman Magnetic Interval in South America, *Geophys. J. R. Astr. Soc.*, **27**, 207–214, 1972.

VALENCIO, D. A., J. F. VILAS, and J. E. MENDÍA, Palaeomagnetism of a sequence of red beds of the Middle and Upper sections of Paganzo Group (Argentina) and the correlation of Upper Palaeozoic-Lower Mesozoic rocks, *Geophys. J. R. Astr. Soc.*, **51**, 59–74, 1977.

AEPS Vol. 1

Hard cover edition to Journal of Geomagnetism and Geoelectricity Vol. 29, No. 4, 1977

Proceedings of AGU 1976 Fall Annual Meeting, December 1976, San Francisco

ORIGIN OF THERMOREMANENT MAGNETIZATION

Edited by David J. DUNLOP

Contents TRM and Its Variation with Grain Size: A Review (R. DAY)/Single Domain Oxide Particles as a Source of Thermoremanent Magnetization (M.E. EVANS) / Domain Structure of Titanomagnetities and Its Variation with Temperature (H.C. SOFFEL) / The Demagnetization Field of Multidomain Grains (R.T. MERRILL) / The Hunting of the 'Psark' (D.J. DUNLOP) / On the Origin of Stable Remanence in Pseudo-Single Domain Grains (S.K. BANERJEE) / The Preparation, Characterization and Magnetic Properties of Synthetic Analogues of Some Carriers of the Palaeomagnetic Record (J.B. O'DONOVAN and W. O'REILLY) / Reduction of Hematite to Magnetite under Natural and Laboratory Conditions (P.N. SHIVE and J.F. DIEHL) / Characteristics of First Order Shock Induced Magnetic Transitions in Iron and Discrimination from TRM (P. WASILEWSKI) / The Thermoremanence Hypothesis and the Origin of Magnetization in Iron Meteorites (A. BRECHER and L. ALBRIGHT) / Thermal Overprinting of Natural Remanent Magnetization and K/Ar Ages in Metamorphic Rocks (K.L. BUCHAN, G.W. BERGER, M.O. MCWILLIAMS, D. YORK, and D.J. DUNLOP) / Does TRM Occur in Oceanic Layer 2 Basalts? (J.M. HALL) / The Effects of Alteration on the Natural Remanent Magnetization of Three Ophiolite Complexes: Possible Implications for the Oceanic Crust (S. LEVI and S.K. BANERJEE)

AEPS Vol. 2

Hard cover edition to Journal of Physics of the Earth Vol. 25, Supplement, 1977 (Not included in regular issues)

Proceedings of the U.S.-Japan Seminar on Theoretical and Experimental Investigations of Earthquake Precursors

EARTHQUAKE PRECURSORS

Edited by C. KISSLINGER and Z. SUZUKI

Contents Earthquake Prediction-Related Research at the Seismological Laboratory, California Institute of Technology, 1974–1976 (J.H. WHITCOMB)/Research on Earthquake Prediction and Related Areas at Columbia University (L.R. SYKES)/Seismic Activities and Crustal Movements the Yamasaki Fault and Surrounding Regions in the Southwest Japan (K. OIKE)/The New Madrid Seismic Zone as a Laboratory for Earthquake Prediction Research (B.J. MITCHELL, W. STAUDER, and C.C. CHENG) / Anomalous Crustal Activity in the Izu Peninsula, Central Honshu (K. TSUMURA) / Recent Seismometrical Works in Japan (S. SUYEHIRO, M. ICHIKAWA, and K. TSUMURA) / Quiet and Violence in Horizontal Movement of the Crust (T. HARADA) / Anomalous Seismic Activity and Earthquake Prediction (H. SEKIYA) / Seismic Activity in the Northeastern Japan Arc (A. TAKAGI, A. HASEGAWA, and N. UMINO) / Observations of Changes in Seismic Wave Velocity in South Kanto District, South of Tokyo, by the Explosion-Seismic Method (T. KAKIMI and I. HASEGAWA) / Some Precursors Prior to Recent Great Earthquakes along the Nankai Trough (H. SATO) / Possibility of Temporal Variations in Earth Tidal Strain Amplitudes Associated with Major Earthquakes (T. MIKUMO, M. KATO, H. DOI, Y. WADA, T. TANAKA, R. SHICHI, and A. YAMAMOTO) / Gravity Changes Associated with Seismic Activities (Y. HAGIWARA) / Geomagnetism in Relation to Tectonic Activities of the Earth's Crust in Japan (N. SUMITOMO) / Precursory and Coseismic Changes in Ground Resistivity (T. RIKITAKE and Y. YAMAZAKI) / Geochemistry as a Tool for Earthquake Prediction (H. WAKITA) / Recent Laboratory Studies of Earthquake Mechanics and Prediction (W.F. BRACE) / Dilatancy of Rocks under General Triaxial Stress States with Special Reference to Earthquake Precursors (K. MOGI) / Possibility of a Great Earthquake in the Tokai District, Central Japan (T. UTSU) / Depth Constraints on Dilatancy Induced Velocity Anomalies (K.W. WINKER and A. NUR) / Seismological Precursors to a Magnitude 5 Earthquake in the Central Aleutian Islands (E.R. ENGDAHL and C. KISSLINGER) / Estimation of Future Destructive Earthquakes from Active Faults on Land in Japan (T. MATSUDA) / Some Problems in the Prediction of the Nemuro-oki Earthquake (K. ABE) / Responses to Earthquake Prediction in Kawasaki City, Japan in 1974 (H. OHTA and K. ABE) / Socioeconomic and Political Consequences of Earthquake Prediction (J.E. HAAS and D.S. MILETI)

AEPS Vol. 3

Proceedings of the U.S.-Japan Seminar on Rare Gas Abundance and Isotopic Constraints on the Origin and Evolution of the Earth's Atmosphere

TERRESTRIAL RARE GASES

Edited by E.C. ALEXANDER, Jr. and M. OZIMA

Contents *EXPERIMENTAL STUDIES* A Mantle Helium Component in Circum-Pacific Volcanic Gases: Hakone, the Marianas, and Mt. Lassen (H. CRAIG, J.E. LUPTON, and Y. HORIBE) / Nitrogen to Argon Ratio in Volcanic Gases (S. MATSUO, M. SUZUKI, and Y. MIZUTANI) / Rare Gas Abundance Pattern of Fumarolic Gases in Japanese Volcanic Areas (O. MATSUBAYASHI, S. MATSUO, I. KANEOKA, and M. OZIMA) / A Review: Some Recent Advances in Isotope Geochemistry of Light Rare Gases (I.N. TOLSTIKHIN) / Abundances and Isotopic Compositions of Rare Gases in Granites and Thucholites (P.K. KURODA and R.D. SHERRILL) / Rare Gas Isotopic Compositions in Diamonds (N. TAKAOKA and M. OZIMA) / Rare Gases in Mantle-Derived Rocks and Minerals (I. KANEOKA, N. TAKAOKA, and K. AOKI) / A Comparison of Terrestrial and Meteoritic Noble Gases (O.K. MANUEL) / The Composition and History of the Martian Atmosphere (T. OWEN) *THEORETICAL STUDIES* Nuclear Components in the Atmosphere (T.J. BERNATOWICZ and F.A. PODOSEK) / Trapped Xenon and Cosmic-Ray Effects in Meteorites, in Lunar Sample, and in the Earth's Materials (K. SAKAMOTO) / Classification and Generation of Terrestrial Rare Gases (K. SAITO) / Earth-Atmosphere Evolution Model Based on Ar Isotopic Data (Y. HAMANO and M. OZIMA) / Terrestrial Potassium and Argon Abundances as Limits to Models of Atmospheric Evolution (D.E. FISHER) / On the Ambient Mantle $^4He/^{40}Ar$ Ratio and the Coherent Model of Degassing of the Earth (D.W. SCHWARTMAN) / Earth Degassing Models, and the Heterogenous vs. Homogeneous Mantle (R. HART and L. HOGAN) / Lead Isotope Constraints on the Early History of the Earth (R.D. RUSSELL) / Matter Accretion into the Solar System (S. HAYAKAWA)

AEPS Vol. 4

Hard cover edition to Journal of Geomagnetism and Geoelectricity Vol. 30, Nos. 3 and 4, 1978

Proceedings of IAGA/IAMAP Joint Assembly, August 1977, Seattle, Washington

AURORAL PROCESSES

Edited by C.T. RUSSELL

Contents *TIMING OF SUBSTORM EVENTS* Pi 2 Micropulsations as Indicators of Substorm Onsets and Intensifications (G. Rostoker and J.V. Olson) / The Use of Ground Magnetograms to Time the Onset of Magnetospheric Substorms (R.L. McPherron) / Substorm Onset in the Magnetotail (A. Nishida) *ELECTROMAGNETIC AND ELECTROSTATIC INSTABILITIES ON AURORAL FIELD LINES* A Review of Electrostatic Wave Measurements on Auroral Magnetic Field Lines (M.C. Kelley) / Diffuse Auroral Precipitation (M. Ashour-Abdalla and C.F. Kennel) / Electromagnetic Plasma Wave Emissions from the Auroral Field Lines (D.A. Gurnett) / Theory of Electromagnetic Waves on Auroral Field Lines (J.E. Maggs) *RAPID AURORAL FLUCTUATIONS AND ASSOCIATED PHENOMENON* Observations of Rapid Auroral Fluctuations (T. Oguti) / Highlights in the Studies of the Relationship of Geomagnetic Field Changes to Auroral Luminosity (W.H. Campbell) / Microburst Precipitation Phenomena (G.K. Parks) *MECHANISMS FOR THE FORMATION OF AURORAL STRUCTURE* Observed Microstructure of Auroral Forms (T.N. Davis) / Birkeland Currents and Auroral Structure (H.R. Anderson) / Relationships between Particle Precipitation and Auroral Forms (J.L. Burch and J.D. Winningham) / Photometric Investigation of Precipitating Particle Dynamics (S.B. Mende) / Generation Mechanisms for Magnetic-Field-Aligned Electric Fields in the Magnetosphere (C.-G. Fälthammar) / Review of Auroral Currents and Auroral Arcs (G. Atkinson) / Acceleration Mechanisms for Auroral Electrons (D.W. Swift) / Subject Index

AEPS Vol. 5

Hard cover edition to Journal of Geomagnetism and Geoelectricity Vol. 30, No. 5, 1978

Proceedings of IAGA/IAMAP Joint Assembly, August 1977, Seattle, Washington

TECTONOMAGNETICS AND LOCAL GEOMAGNETIC FIELD VARIATIONS

Edited by M. Fuller, M.J.S. Johnston, and T. Yukutake

Contents Symposium on Tectonomagnetics and Small Scale Secular Variations Held at the IAGA/IAMAP Joint Assembly at Seattle on Tuesday, August 22nd, 1977 (V.A. Shapiro and M.J.S. Johnston) / Tectonomagnetic Studies in Tajikstan (Yu. P. Skovorodkin, L.S. Bezuglaya, and T.V. Guseva) / An Attempt to Observe a Seismomagnetic Effect during the Gazly 17th May 1976 Earthquake (V.A. Shapiro and K.N. Abdullabekov) / Secular Variation Anomalies and Aseismic Geodynamic in the Urals (V.A. Shapiro, A.L. Aleinkov, A.A. Nulman, V.A. Pyankov, and A.V. Zubkov) / Geomagnetic Investigations in the Seismoactive Regions of Middle Asia (V.A. Shapiro, A.N. Pushkov, K.N. Abdullabekov, E.B. Berdaliev, and M.Yu. Muminov) / Local Magnetic Field Variations and Stress Changes Near a Slip Discontinuity on the San Andreas Fault (M.J.S. Johnston) / Geomagnetic Secular Variation Anomalies in the GDR (W. Mundt) / Noise Reduction Techniques for Use in Determining Local Geomagnetic Field Changes (R.H. Ware and P.L. Bender) / Local Variations in Magnetic Field, Long-Term Changes in Creep Rate, and Local Earthquakes along the San Andreas Fault in Central California (B.E. Smith, M.J.S. Johnston, and R·O. Burford) / Geomagnetic Induction Study of the Seismically Active Fault along the Southwestern Coast of the Sea of Japan (J. Miyakoshi and A. Suzuki) / Time Dependence of Magnetotelluric Fields in a Tectonically Active Region in Eastern Canada (R.D. Kurtz and E.R. Niblett) / Piezomagnetic Response with Depth, Related to Tectonomagnetism as an Earthquake Precursor (R.S. Carmichael) / Magnetic Susceptibility of Magnetite under Hydrostatic Pressure, and Implications for Tectonomagnetism (A.A. Nulman, V.A. Shapiro, S.I. Maksimovskikh, N.A. Ivanov, J. Kim, and R.S. Carmichael) / Effect of Uniaxial Stress upon Remanent

Magnetization: Stress Cycling and Domain State Dependence (J. REVOL, R. DAY, and M. FULLER) / On the Measurement of Stress Sensitivity of NRM Using a Cryogenic Magnetometer (T.L. HENYEY, S.J. PIKE, and D.F. PALMER)

AEPS Vol. 6

Hard cover edition to Journal of Physics of the Earth Vol. 26, Supplement, 1978 (Not included in regular issues)

Proceedings of the International Conference on Geodynamics of the Western Pacific-Indonesian Region, March 1978, Tokyo

GEODYNAMICS OF THE WESTERN PACIFIC

Edited by S. UYEDA, R.W. MURPHY, and K. KOBAYASHI

Contents Plate Tectonic Evolution of North Pacific Rim (W.R. DICKINSON) / Speculations on Mountain Building and the Lost Pacific Continent (A. NUR and Z. BEN-AVRAHAM) / Benioff Zones, Absolute Motion and Interarc Basin (F.T. WU) / Oceanic Crust in the Dynamics of Plate Motion and Back-Arc Spreading (Y. IDA) / Basic Types of Internal Deformation of the Continental Plate at Arc-Arc Junctions (K. SHIMAZAKI, T. KATO, and K. YAMASHINA) / Fault Patterns in Outer Trench Walls and Their Tectonic Significance (G.M. JONES, T.W.C. HILDE, G.F. SHARMAN, and D.C. AGNEW) / Motion of the Pacific Plate and Formation of Marginal Basins: Asymmetric Flow Induction (R.C. BOSTROM) / The Relationship between Volcanic Island Genesis and the Indo-Australian Pacific Plate Margins in the Eastern Outer Islands, Solomon Islands, South-West Pacific (G. Wyn HUGHES) / Upper Mantle Velocity Structure in the New Hebrides Island Arc Region (K.L. KAILA and V.G. KRISHNA) / Upper Mantle Velocity Structure in the Tonga-Kermadec Island Arc Region (K.L. KAILA and V.G. KRISHNA) / Morphology and Structure of the Southern Part of the New Hebrides Island Arc System (J. DANIEL) / Paleomagnetic Evidence for the Rotation of Seram, Indonesia (N.S. HAILE) / A Late Miocene K-Ar Age for the Lavas of Pulau Kelang, Seram, Indonesia (R.D. BECKINSALE and S. NAKAPADUNGRAT) / A Survey of Paleomagnetic Data on Mexico (S. PAL) / Southeast Asian Tin Granitoids of Contrasting Tectonic Setting (C.S. HUTCHISON) / Seismicity, Gravity and Tectonics in the Andaman Sea (R.K. VERMA, M. MUKHOPADHYAY, and N.C. BHUIN) / Focal Mechanisms and Tectonics in the Taiwan-Philippine Region (T. SENO and K. KURITA) / Recent Tectonics of Taiwan (F.T. WU) / Tectonics of the Ryukyu Island Arc (K. KIZAKI) / Explosion Seismic Studies in South Kyushu Especially around the Sakurajima Volcano (K, ONO, K. ITO, I. HASEGAWA, K. ICHIKAWA, S. IIZUKA, T. KAKUTA, and H. SUZUKI) / Two Types of Accretionary Fold Belts in Central Japan (Y. OGAWA and K. HORIUCHI) / Permain and Triassic Sedimentary History of the Honshu Geosyncline in the Tamba Belt, Southwest Japan (D. SHIMIZU, N. IMOTO, and M. MUSASHINO) / Thermal Structure of the Sanbagawa Metamorphic Belt in Central Shikoku (S. BANNO, T. HIGASHINO, M. OTSUKI, T. ITAYA, and T. NAKAJIMA) / Shimanto Geosyncline and Kuroshio Paleoland (T. HARATA, K. HISATOMI, F. KUMON, K. NAKAZAWA, M. TATEISHI, H. SUZUKI, and T. TOKUOKA) / Magnetic Stratigraphy of the Japanese Neogene and the Development of the Island Arcs of Japan (N. NIITSUMA) / Regional Characteristics and Their Geodynamic Implications of Late Quaternary Tectonic Movement Deduced from Deformed Former Shorelines in Japan (Y. OTA and T. YOSHIKAWA) / Magnetic Anomalies and Tectonic Evolution of the Shikoku Inter-Arc Basin (K. KOBAYASHI and M. NAKADA) / A Compilation of Magnetic Data in the Northwestern Pacific and in the North Philippine Sea (N. ISEZAKI and H. MIKI) / Collision of the Izu-Bonin Arc with Central Honshu: Cenozoic Tectonics of the Fossa Magna, Japan (T. MATSUDA) / Flow under the Island Arc of Japan and Lateral Variation of Magma Chemistry of Island Arc Volcanoes (M. TORIUMI) / Seismic Activity and Pore Pressures across Island Arcs of Japan (N. FUJII and K. KURITA) / Aseismic Belt along the Frontal Arc and Plate Subduction in Japan (K. YAMASHINA, K. SHIMAZAKI, and T. KATO) / Tsunamicity of Sanriku Depends on Subduction Tectonics (Wm. M. ADAMS) / Seismic Studies of the Upper Mantle beneath the Arc-Junction at Hokkaido: Folded Structure of Intermediate-Depth Seismic Zone and Altenuation of Seismic Waves (T. MORIYA) / Sedimentary Patterns in Apparent Back-Arc Basins: A Case Study of the Neogene Sequence in Northwestern Hokkaido, Japan (H. OKADA) / Velocity Anisotropy in the Sea of Japan as Revealed by Big Explosions (H. OKADA, T. MORIYA, T. MASUDA, T. HASEGAWA, S. ASANO, K. KASAHARA, A. IKAMI, H. AOKI, Y. SASAKI, N. HURUKAWA, and K. MATSUMURA) / Geodynamics of the North-Eastern Asia in Mesozoic and Cenozoic Time and the Nature of Volcanic Belts (L.M. PARFENOV, I.P. VOINOVA, B.A. NATAL'IN, and D.F. SEMENOV) / The Crustal Structure and Origin of the Basins of Japan Sea and Some Other Seas of the Circum-Pacific Mobile Belt (P.N. KROPOTKIN) / Major Strike-Slip Faults and Their Bearing on Spreading in the Japan Sea (K. OTSUKI and M. EHIRO) / Significant Eruptive Activities Related to Large Interplate Earthquakes in the Northwestern Pacific Margin (M. KIMURA) / A Mechanism to Explain the Earthquakes around Japan by the Process of Partial Melting (M. HAYAKAWA and S. IIZUKA) / The Formation of Intermediate and Deep Earthquake Zone in Relation to the Geologic Development of East Asia since Mesozoic (Y. SUZUKI, K. KODAMA, and T. MITSUNASHI) / Subject Index / Geographical Index

AEPS Vol. 7

Hard cover edition to Journal of Geomagnetism and Geoelectricity Vol. 31, No. 3, 1979

Proceedings of IAGA/IAMAP Joint Assembly August 1977, Seattle, Washington

ELECTRIC CURRENT AND ATMOSPHERIC MOTION.

Edited by S. KATO and R.G. ROPER

Contents Electrodynamics of the Ionosphere from Incoherent Scatter: A Review (M. BLANC)/ Long-Period Waves in Mesospheric Winds at Saskatoon (52°N) (A.D. BELMONT and G.D. NASTROM)/Solar Tidal Wind Structures and the *E*-Region Dynamo (J.M. FORBES and H.B. GARRETT)/Dynamics of Severe Storms through the Study of Thermospheric-Tropospheric Coupling (R.J. HUNG and R.E. SMITH)/Coordinated Measurements of *E*-Layer Drifts (E.S. KAZIMIROVSKY and V.D. KOKOUROV)/On an Origin of Ultra Long Period (Several Days) of Geomagnetic Fluctuations (T. KITAMURA)/IMF and Lower Thermospheric Currents and Motions: A Review (S. MATSUSHITA)/Abnormal Features of the Regular Daily Variation S_R (P.N. MAYAUD)/Rocket Measurements of Annual Mean Prevailing, Diurnal and Semi-Diurnal Winds in the Lower Thermosphere at Mid-Latitudes (D. REES)/Mid-Latitude Winds and Electric Fields in the Lower Thermosphere and Their Relationship with the Global S_q Ionospheric Current System (D. REES)/Ionospheric Wind Dynamo Theory: A Review (A.D. RICHMOND)/The Quiet-time Equatorial Electrojet and Counter-Electrojet (R.T. MARRIOTT, A.D. RICHMOND, and S.V. VENKATESWARAN)/Equatorial Electrojet and S_q Current System —Part I (J. W. MACDOUGALL)/Equatorial Electrojet and S_q Current System—Part II (J.W. MACDOUGALL)/Results from *in situ* Measurements of Ionospheric Currents in the Equatorial Region—I (S. SAMPATH and T.S.G. SASTRY)/Depth of Non-Conducting Layer in the Indian Ocean Region around Thumba, Derived from *in situ* investigations of Equatorial Electrojet—II (S. SAMPATH and T.S.G. SASTRY)/AC Electric Fields Associated with the Plasma Instabilities in the Equatorial Electrojet—III (S. SAMPATH and T.S.G. SASTRY)/Electric Potential

Difference between Conjugate Points in Middle Latitudes Caused by Asymmetric Dynamo in the Ionosphere (N. FUKUSHIMA) / Results of Wind Velocity Measurements at Middle and High Latitudes by the Meteor Radar Method (I.A. LYSENKO, A.D. ORLYANSKY, and Yu. I. PORTNYAGIN) / A Comparison between Radio Meteor and Airglow Winds (G. HERNANDEZ and R.G. ROPER) / Subject Index

AEPS Vol. 8

Hard cover edition to Journal of Physics of the Earth Vol. 27, Supplement, 1979 (Not included in regular issues)

STRUCTURE OF TRANSITION ZONE

Edited by S. ASANO

Contents Crust and Upper Mantle Structure beneath Northeastern Honshu, Japan as Derived from Explosion Seismic Observations (S. ASANO, H. OKADA, T. YOSHII, K. YAMAMOTO, T. HASEGAWA, K. ITO, S. SUZUKI, A. IKAMI, and K. HAMADA) / Regionality of the Upper Mantle around Northeastern Japan as Revealed by Big Explosions at Sea. I. SEIHA-1 Explosion Experiment (H. OKADA, S. ASANO, T. YOSHII, A. IKAMI, S. SUZUKI, T. HASEGAWA, K. YAMAMOTO, K. ITO, and K. HAMADA) / On the Junction Character of the Continental and the Oceanic Lithospheric Blocks in the Kamchatka Region (G.I. ANOSOV, S.K. BIKKENINA, V.I. FEDORCHENKO, A.A. POPOV, K.F. SERGEEV, and V.K. UTNASIN) / New Evidences of the Discontinuous Structure of the Descending Lithosphere as Revealed by *ScSp* Phase (Hm. OKADA)/ A Block Velocity Model of the Focal Zone and Adjacent Mantle in the Kurile-Japan Region (R.Z. TARAKANOV) / Geological Structure of the Southwestern Okhotsk Sea Area (S.L. SOLOVIEV, M.L. KRASNY, O.A. MELNIKOV, Yu.A. PAVLOV, E.I. POPOV, S.S. SNEGOVSKOY, I.K. TUENOV, and B.I. VASILIEV) / Structure and Geological Nature of the Kuril Abyssal Basin in the Okhotsk Sea (I.K. TUEZOV, B.I. VASILIEV, M.L. KRASNY, Yu.A. PAVLOV, and E.I. POPOV) / The Daito Ridge Group and the Kyushu-Palau Ridge—with Special Reference to the Tectonics of the Philippine Sea— (T. SHIKI, Y. MISAWA, and I. KONDA) / Heat Flow in the Hokkaido-Okhotsk Region and Its Tectonic Implications (S. EHARA) / Geothermal Investigations Carried out in the Northwestern Sector of the Pacific Mobile Belt (O.V. VESELOV, N.A. VOLKOVA, V.V. SOINOV, and Y.D. EREMIN) / Deep Electric Conductivity Study in the Asia-to-Pacific Transition Zone (L.L. VANYAN, V.V. BORETZ, A.M. LYAPISHEV, B.E. MARDERFELD, A.V. RODIONOV, and V.N. VERKHOVSKY) / Electrical Conductivity Structure beneath the Japan Island Arc by Geomagnetic Induction Study (J. MIYAKOSHI) / Geomagnetic Anomalies of the Japan Sea by Japanese and Russian Magnetic Data (N. ISEZAKI, M. YASUI, and S. UYEDA) / Recent Crustal Movements in Primorie and Sakhalin (V.K. ZAKHAROV, B.A. BELOUSOV, V.P. SEMAKIN, N.F. VASILENKO, and G.G. YAKUSHKO) / Subject Index

AEPS Vol. 9

Hard cover edition to Journal of Geomagnetism and Geoelectricity, Vol. 32, Supplement I, 1980 (Not included in regular issues)

ELECTROMAGNETIC INDUCTION IN THE EARTH AND MOON

Edited by U. SCHMUCKER

Contents Observation of Very Low Frequency Electromagnetic Signals in the Ocean (J. H. FILLOUX)/Atlantic Lithosphere Sounding (C. S. COX, J. H. FILLOUX, D. I. GOUGH, J. C. LARSEN, K. A. POEHLS, R. P. VON HERZEN, and R. WINTER)/North Pacific Magnetotelluric Experiments (J. H.

FILLOUX)/Electromagnetic Induction in the Vancouver Island Region (W. NIENABER, H. W. DOSSO, L. K. LAW, F. W. JONES, and V. RAMASWAMY)/Induction in Arbitrarily Shaped Oceans II: Edge Correction for the Case of Infinite Conductivity (R. C. HEWSON-BROWNE and P. C. KENDALL)/The Effect of a Simple Model of the Pacific Ocean on S_q Variations (B. A. HOBBS and G. J. K. DAWES)/Electromagnetic Induction at a Model Ocean Coast (G. FISCHER, P.-A. SCHNEGG, and K. D. USADEL)/Diakoptic Solution of Induction Problems (C. R. BREWITT-TAYLOR and P. B. JOHNS)/ Induction by S_q (W. D. PARKINSON)/Electromagnetic Response Functions from Interrupted and Noisy Data (J. C. LARSEN)/Deep Conductivity Distribution on the Russian Platform from the Results of Combined Magnetotelluric and Global Magnetovariational Data Interpretation (A. A. KOVTUN and L. N. POROKHOVA)/Connection between the Electric Conductivity Increase due to Phase Transition and Heat Flow (A. ÁDÁM)/Geomagnetic Variations Behavior in Central Europe (I. I. ROKITYANSKY)/Geomagnetic Sounding of an Ancient Plate Margin in the Canadian Appalachians (J. A. WRIGHT and N. A. COCHRANE)/Magnetovariational and Magnetotelluric Investigations in S. Scotland (V. R. S. HUTTON and A. G. JONES)/An Analogue Model Study of Ocean-Wave Induced Magnetic Field Variations Near a Coastline (T. MILES and H. W. DOSSO)/Long Period Variations of the Geomagnetic Field and Inferences about the Deep Electric Conductivity of the Earth (A. M. IŞIKARA)/Inverse Magnetotelluric Problem for Sounding Curves from Czechoslovak Localities (K. PĚČ, J. PĚČOVÁ, and O. PRAUS)

GPSR Compliance

The European Union's (EU) General Product Safety Regulation (GPSR) is a set of rules that requires consumer products to be safe and our obligations to ensure this.

If you have any concerns about our products, you can contact us on ProductSafety@springernature.com

In case Publisher is established outside the EU, the EU authorized representative is:

Springer Nature Customer Service Center GmbH
Europaplatz 3
69115 Heidelberg, Germany

Batch number: 10585152

Printed by Printforce, the Netherlands